Renewable Energy Unleashed

Strategies for a Cleaner, Greener Environment

Lara Willis

1

Renewable Energy Unleashed

TABLE OF CONTENTS

Chapter 1: Solar Power: Illuminating the Path Forward

Basics of Solar Energy Technology

Solar energy technology has emerged as a cornerstone of the renewable energy landscape, offering a sustainable solution to the world's growing energy demands. At its core, solar energy harnesses the power of the sun, converting sunlight into electricity through a process that is both efficient and environmentally friendly. This chapter delves into the fundamental aspects of solar energy technology, providing a comprehensive understanding of its mechanisms, applications, and potential.

The journey of solar energy begins with photovoltaic (PV) cells, the building blocks of solar panels. These cells are made from semiconductor materials, typically silicon, which exhibit the photovoltaic effect. When sunlight strikes the surface of a PV cell, it excites electrons, creating an electric current. This direct conversion of light into electricity is what makes solar technology so appealing. The efficiency of a solar panel is determined by its ability to convert sunlight into usable electricity, and advancements in materials and design have steadily improved this efficiency over the years.

Solar panels are typically grouped into arrays, which can be installed on rooftops, in open fields, or integrated into building materials. The versatility of solar installations allows for a wide range of applications, from small-scale residential systems to large-scale solar farms. Residential solar systems are designed to meet the energy needs of individual households, often supplemented by battery storage solutions to provide power during cloudy days or at night. Commercial solar installations, on the other hand, are larger in scale and can power entire buildings or industrial facilities, contributing significantly to reducing carbon footprints.

One of the most exciting developments in solar energy technology is the innovation in solar energy storage. Traditional solar systems are limited by their reliance on sunlight, but advancements in battery technology have paved the way for more reliable and consistent energy supply. Lithium-ion batteries, for example, have become a popular choice for storing solar energy due to their high energy density and long lifespan. These batteries allow excess energy generated during peak sunlight hours to be stored and used when needed, enhancing the overall efficiency and reliability of solar systems.

The economic and environmental advantages of solar energy are compelling. Solar power is a clean, renewable resource that produces no greenhouse gas emissions during operation. This makes it an attractive option for reducing the environmental impact of energy production. Economically, the cost of solar panels has decreased significantly over the past decade, making solar energy more accessible to a broader audience. Government incentives and subsidies further encourage the adoption of solar technology, providing financial benefits to those who invest in solar systems.

Despite its many advantages, the implementation of solar energy technology is not without challenges. One of the primary obstacles is the initial cost of installation, which can be prohibitive for some individuals and businesses. However, the long-term savings on energy bills and the potential for government incentives often offset these initial expenses. Another challenge is the intermittent nature of solar energy, as it is dependent on weather conditions and daylight hours. This limitation can be mitigated by integrating solar systems with other renewable energy sources or by utilizing advanced energy storage solutions.

The integration of solar energy into existing energy grids also presents technical challenges. Solar power generation is decentralized, meaning it is produced at multiple locations rather than a single power plant. This requires sophisticated grid management systems to ensure a stable and reliable energy supply. Smart grid technology, which uses digital communication to monitor and manage energy flow, is essential for optimizing the integration of solar energy into national grids.

In addition to these technical challenges, there are also regulatory and policy considerations that impact the adoption of solar energy technology. Governments play a crucial role in shaping the renewable energy landscape through policies that promote or hinder the development of solar infrastructure. Supportive policies, such as feed-in tariffs and tax credits, can accelerate the growth of solar energy by providing financial incentives for both producers and consumers.

The future of solar energy technology is bright, with ongoing research and development driving innovation and efficiency. Emerging technologies, such as perovskite solar cells and bifacial panels, promise to further enhance the performance and affordability of solar systems. Perovskite cells, for instance, offer the potential for higher efficiency and lower production costs compared to traditional silicon-based cells. Bifacial panels, which capture sunlight on both sides, can increase energy output by utilizing reflected light from the ground.

As solar technology continues to evolve, it is poised to play an increasingly important role in the global transition to renewable energy. The potential for solar energy to provide a sustainable and reliable source of power is immense, and its widespread adoption could significantly reduce the world's reliance on fossil fuels. By embracing solar energy technology, we can pave the way for a cleaner, more sustainable future, harnessing the power of the sun to illuminate the path forward.

Residential vs. Commercial Solar Solutions

Solar energy has become a pivotal player in the quest for sustainable energy solutions, and its applications span both residential and commercial sectors. Understanding the differences and similarities between these two applications is crucial for anyone considering a transition to solar power. Each sector presents unique opportunities and challenges, shaped by factors such as scale, cost, and energy needs.

Residential solar solutions are designed to cater to the energy requirements of individual households. These systems typically involve the installation of solar panels on rooftops, where they can capture sunlight and convert it into electricity. The primary goal of residential solar is to reduce or eliminate a household's reliance on grid electricity, thereby lowering energy bills and reducing carbon footprints. Homeowners often choose solar systems based on their energy consumption patterns, available roof space, and budget constraints.

One of the key advantages of residential solar systems is their ability to provide energy independence. By generating their own electricity, homeowners can shield themselves from fluctuating energy prices and potential grid outages. Additionally, residential solar installations can increase property values, as more buyers are attracted to homes with sustainable energy solutions. The financial incentives offered by governments, such as tax credits and rebates, further enhance the appeal of residential solar systems.

However, the implementation of residential solar solutions is not without challenges. The initial cost of purchasing and installing solar panels can be significant, although this is often offset by long-term savings on energy bills. Homeowners must also consider the suitability of their property for solar installation, as factors like roof orientation, shading, and local climate can impact the efficiency of solar panels. Moreover, navigating the various financing options and incentive programs can be complex, requiring careful research and planning.

In contrast, commercial solar solutions are tailored to meet the energy needs of businesses and industrial facilities. These systems are typically larger in scale and may involve ground-mounted solar arrays or expansive rooftop installations. The primary objective of commercial solar is to reduce operational costs by generating a significant portion of a business's energy needs on-site. This not only lowers electricity bills but also enhances a company's sustainability profile, which can be a valuable asset in today's environmentally conscious market.

Commercial solar installations offer several benefits beyond cost savings. By adopting solar energy, businesses can demonstrate their commitment to sustainability, which can improve brand reputation and customer loyalty. Additionally, commercial solar systems can provide a hedge against energy price volatility, offering businesses greater financial stability. For companies with large energy demands, solar power can also contribute to meeting corporate sustainability goals and regulatory requirements.

Despite these advantages, commercial solar solutions face their own set of challenges. The scale of these projects often requires significant upfront investment, and businesses must carefully evaluate the return on investment. Site selection and system design are critical, as factors such as land availability, shading, and local regulations can influence the feasibility and efficiency of a solar installation. Furthermore, businesses must consider the potential impact on operations during installation and maintenance, as well as the integration of solar power with existing energy systems.

Both residential and commercial solar solutions share common elements, such as the use of photovoltaic technology and the potential for energy storage. However, the scale and complexity of commercial systems often necessitate a more comprehensive approach to design, financing, and implementation. Businesses may need to engage with solar developers, financial institutions, and regulatory bodies to successfully navigate the transition to solar energy.

The decision to adopt solar energy, whether for residential or commercial purposes, involves careful consideration of various factors. For homeowners, the focus is often on achieving energy independence and reducing household expenses. For businesses, the emphasis is on cost savings, sustainability, and enhancing corporate image. In both cases, the benefits of solar energy extend beyond financial gains, contributing to a cleaner and more sustainable future.

As solar technology continues to advance, the distinction between residential and commercial solutions may become less pronounced. Innovations in solar panel efficiency, energy storage, and system integration are making solar energy more accessible and cost-effective for all users. Whether for a single-family home or a large industrial complex, solar energy offers a viable path toward reducing reliance on fossil fuels and mitigating the impacts of climate change.

Ultimately, the choice between residential and commercial solar solutions depends on individual needs, goals, and circumstances. By understanding the unique characteristics and considerations of each sector, individuals and businesses can make informed decisions that align with their energy objectives and contribute to a more sustainable world.

Innovations in Solar Energy Storage

Solar energy storage has become a pivotal aspect of the renewable energy landscape, addressing one of the most significant challenges associated with solar power: its intermittent nature. As the sun does not shine consistently throughout the day, and certainly not at night, the ability to store solar energy for later use is crucial for maximizing the efficiency and reliability of solar systems. Innovations in solar energy storage are transforming the way we harness and utilize solar power, making it a more viable and attractive option for both residential and commercial applications.

At the heart of solar energy storage lies the development of advanced battery technologies. Traditional lead-acid batteries, once the mainstay of energy storage, are gradually being replaced by more efficient and durable alternatives. Lithium-ion batteries have emerged as a leading choice due to their high energy density, long lifespan, and declining costs. These batteries are capable of storing large amounts of energy in a compact form, making them ideal for both small-scale residential systems and large-scale commercial installations.

One of the most exciting advancements in solar energy storage is the advent of solid-state batteries. Unlike conventional batteries that use liquid electrolytes, solid-state batteries employ solid electrolytes, which offer several advantages. They are safer, as they eliminate the risk of leakage and thermal runaway, and they can potentially offer higher energy densities and faster charging times. Although still in the developmental stage, solid-state batteries hold the promise of revolutionizing solar energy storage by providing more efficient and reliable solutions.

Flow batteries represent another innovative approach to solar energy storage. These batteries store energy in liquid electrolytes contained in external tanks, allowing for easy scalability. The capacity of a flow battery can be increased simply by adding more electrolyte, making it an attractive option for large-scale solar installations. Flow batteries also boast long cycle lives and can be discharged and recharged thousands of times without significant degradation, offering a durable and flexible storage solution.

In addition to advancements in battery technology, other innovative storage solutions are gaining traction. Thermal energy storage, for example, involves capturing and storing heat generated by solar power for later use. This can be particularly useful in applications such as heating water or providing space heating in buildings. Thermal storage systems can be integrated with solar thermal collectors, which convert sunlight into heat, offering an efficient way to store and utilize solar energy.

Hydrogen storage is another promising avenue for solar energy storage. By using solar power to electrolyze water, hydrogen gas can be produced and stored for later use. This hydrogen can then be converted back into electricity through fuel cells or used directly as a clean fuel source. Hydrogen storage offers the advantage of long-term energy storage, as hydrogen can be stored indefinitely without loss of energy. This makes it an attractive option for balancing seasonal variations in solar energy production.

The integration of smart grid technology with solar energy storage systems is also driving innovation. Smart grids use digital communication and automation to optimize the distribution and consumption of electricity. By integrating solar storage systems with smart grids, energy can be managed more efficiently, ensuring that stored solar power is used when it is most needed. This not only enhances the reliability of solar systems but also contributes to grid stability and reduces the need for fossil fuel-based backup power.

Innovations in solar energy storage are not limited to technological advancements; they also encompass new business models and financing options. Energy-as-a-service (EaaS) is an emerging model that allows consumers to access solar energy storage solutions without the need for upfront investment. Under this model, service providers install and maintain solar storage systems, and consumers pay for the energy they use. This approach lowers the barrier to entry for solar energy storage, making it more accessible to a wider audience.

The development of community solar projects is another innovative approach to solar energy storage. These projects allow multiple users to share the benefits of a single solar installation, often combined with energy storage. Community solar projects can provide energy to those who may not have suitable rooftops for solar panels or who cannot afford individual systems. By pooling resources, participants can enjoy the benefits of solar energy and storage at a reduced cost, fostering greater adoption of renewable energy.

As solar energy storage technology continues to evolve, it is poised to play an increasingly important role in the transition to renewable energy. The ability to store and manage solar power effectively is crucial for overcoming the limitations of solar energy and ensuring a reliable and sustainable energy supply. By embracing these innovations, we can unlock the full potential of solar energy, paving the way for a cleaner and more sustainable future.

Economic and Environmental Advantages

Solar energy stands as a beacon of hope in the quest for sustainable and environmentally friendly energy solutions. Its economic and environmental advantages are compelling, making it an attractive option for individuals, businesses, and governments alike. By harnessing the power of the sun, solar energy offers a pathway to reducing carbon emissions, lowering energy costs, and fostering economic growth.

One of the most significant economic advantages of solar energy is its potential to reduce electricity bills. For homeowners and businesses, installing solar panels can lead to substantial savings on energy costs. Once the initial investment in solar technology is made, the ongoing cost of generating electricity is minimal. Solar panels require little maintenance and have no fuel costs, as they rely solely on sunlight. This translates to long-term savings, as users can generate their own electricity and reduce their reliance on grid power.

The decreasing cost of solar technology has further enhanced its economic appeal. Over the past decade, the price of solar panels has dropped significantly, making solar energy more accessible to a broader audience. Technological advancements and economies of scale have contributed to this decline, allowing more people to invest in solar systems. Additionally, government incentives, such as tax credits and rebates, have made solar energy even more affordable, encouraging widespread adoption.

Solar energy also offers economic benefits at a macro level. The growth of the solar industry has created numerous jobs in manufacturing, installation, and maintenance. As demand for solar technology increases, so does the need for skilled workers, contributing to job creation and economic development. The solar industry has become a significant driver of employment, providing opportunities for workers in both urban and rural areas.

From an environmental perspective, solar energy is a clean and renewable resource that produces no greenhouse gas emissions during operation. This makes it a powerful tool in the fight against climate change. By replacing fossil fuel-based energy sources with solar power, we can significantly reduce carbon emissions and mitigate the impacts of global warming. The environmental benefits of solar energy extend beyond carbon reduction, as it also helps decrease air and water pollution associated with traditional energy production.

Solar energy's minimal environmental footprint is another advantage. Unlike fossil fuel power plants, which require large amounts of water for cooling, solar panels use no water during operation. This reduces the strain on water resources, particularly in arid regions where water scarcity is a concern. Furthermore, solar installations have a relatively small land footprint compared to other renewable energy sources, such as wind or hydropower, making them suitable for a variety of locations.

The integration of solar energy into existing energy systems can also enhance grid stability and resilience. By decentralizing energy production, solar power reduces the risk of large-scale power outages and increases the reliability of energy supply. Distributed solar systems, such as rooftop installations, can provide localized energy generation, reducing the need for long-distance transmission and minimizing energy losses. This not only improves the efficiency of the energy system but also enhances its resilience to natural disasters and other disruptions.

Solar energy's ability to provide energy independence is another significant advantage. By generating their own electricity, individuals and businesses can reduce their reliance on external energy sources and protect themselves from fluctuating energy prices. This energy independence can be particularly beneficial in remote or off-grid areas, where access to traditional energy sources may be limited or unreliable. Solar energy offers a sustainable and reliable solution for meeting energy needs in these regions.

The economic and environmental advantages of solar energy are closely intertwined, creating a positive feedback loop that drives further adoption and innovation. As more people and businesses invest in solar technology, the demand for solar products and services increases, spurring further advancements and cost reductions. This, in turn, makes solar energy even more attractive, encouraging more widespread use and contributing to a cleaner and more sustainable future.

In conclusion, solar energy presents a compelling case for both economic and environmental benefits. Its ability to reduce energy costs, create jobs, and decrease carbon emissions makes it a vital component of the global transition to renewable energy. By embracing solar power, we can pave the way for a more sustainable and prosperous future, harnessing the power of the sun to drive economic growth and protect our planet for generations to come.

Overcoming Implementation Challenges

Implementing solar energy solutions, while promising in terms of sustainability and cost savings, presents a unique set of challenges that must be addressed to ensure successful adoption. These challenges range from technical and financial hurdles to regulatory and social barriers. Understanding and overcoming these obstacles is crucial for individuals, businesses, and policymakers seeking to harness the full potential of solar energy.

One of the primary challenges in implementing solar energy systems is the initial cost of installation. Although the price of solar panels has decreased significantly over the years, the upfront investment required for purchasing and installing a solar system can still be substantial. This financial barrier can deter potential adopters, particularly in residential settings where budget constraints are more pronounced. To address this issue, various financing options have emerged, such as solar loans, leases, and power purchase agreements (PPAs). These options allow individuals and businesses to spread the cost of solar installations over time, making them more affordable and accessible.

Another significant challenge is the variability of solar energy production. Solar power generation is inherently dependent on weather conditions and daylight hours, leading to fluctuations in energy output. This intermittency can pose a problem for grid stability and reliability, particularly in regions with high solar penetration. To mitigate this issue, energy storage solutions, such as batteries, are increasingly being integrated with solar systems. These storage technologies allow excess energy generated during peak sunlight hours to be stored and used when needed, smoothing out the variability and ensuring a consistent energy supply.

The integration of solar energy into existing energy grids also presents technical challenges. Solar power generation is decentralized, meaning it is produced at multiple locations rather than a single power plant. This requires sophisticated grid management systems to ensure a stable and reliable energy supply. Smart grid technology, which uses digital communication to monitor and manage energy flow, is essential for optimizing the integration of solar energy into national grids. Additionally, grid infrastructure may need to be upgraded to accommodate the increased flow of electricity from distributed solar systems.

Regulatory and policy barriers can also hinder the implementation of solar energy solutions. In some regions, outdated regulations and bureaucratic red tape can slow down the permitting and installation process, discouraging potential adopters. Streamlining these processes and implementing supportive policies, such as feed-in tariffs and tax incentives, can accelerate the growth of solar energy by providing financial benefits and reducing administrative burdens. Policymakers play a crucial role in shaping the renewable energy landscape, and their support is vital for overcoming regulatory challenges.

Social acceptance and awareness are other factors that can impact the implementation of solar energy systems. Public perception of solar energy can influence its adoption, and misconceptions about its reliability, cost, and environmental impact can create resistance. Education and outreach efforts are essential for raising awareness and dispelling myths about solar energy. By providing accurate information and highlighting the benefits of solar power, stakeholders can foster greater acceptance and encourage more widespread adoption.

Site selection and land use considerations are additional challenges that must be addressed when implementing solar energy systems. For large-scale solar installations, finding suitable land that is free from shading and has optimal sun exposure is critical. Land use conflicts can arise, particularly in densely populated areas or regions with competing land interests. Careful planning and stakeholder engagement are necessary to identify appropriate sites and minimize potential conflicts. In some cases, innovative solutions, such as floating solar panels on bodies of water or integrating solar panels into agricultural land, can provide alternative options for site selection.

The rapid pace of technological advancement in the solar industry can also present challenges for implementation. As new technologies and products are developed, keeping up with the latest innovations can be daunting for consumers and businesses. Staying informed about the latest trends and advancements is crucial for making informed decisions about solar energy investments. Engaging with industry experts, attending conferences, and participating in training programs can help stakeholders stay abreast of the latest developments and ensure they are implementing the most efficient and effective solutions.

Finally, the long-term maintenance and performance monitoring of solar energy systems are essential for ensuring their continued success. Regular maintenance is necessary to keep solar panels and associated equipment in optimal condition, and performance monitoring can help identify and address any issues that may arise. Establishing a maintenance plan and working with qualified professionals can help ensure the longevity and efficiency of solar installations.

By addressing these challenges and implementing effective strategies, individuals, businesses, and policymakers can overcome the barriers to solar energy adoption. The successful implementation of solar energy solutions requires a comprehensive approach that considers technical, financial, regulatory, and social factors. By working collaboratively and leveraging available resources and expertise, stakeholders can unlock the full potential of solar energy and contribute to a more sustainable and resilient energy future.

Chapter 2: Harnessing Wind Energy: A Breath of Fresh Air

Wind Turbine Design and Functionality

Wind turbines have become iconic symbols of renewable energy, capturing the power of the wind to generate electricity in a clean and sustainable manner. Understanding the design and functionality of these towering structures is essential for appreciating their role in the global energy landscape. Wind turbines are marvels of engineering, combining aerodynamic principles with advanced technology to convert kinetic energy from the wind into electrical power.

At the heart of a wind turbine's design is its rotor, which consists of multiple blades attached to a central hub. These blades are typically made from lightweight, durable materials such as fiberglass or carbon fiber, designed to withstand the stresses of high-speed rotation. The shape of the blades is crucial, as it determines the turbine's efficiency in capturing wind energy. They are carefully crafted with an aerodynamic profile, similar to an airplane wing, to maximize lift and minimize drag. As the wind flows over the blades, it creates a pressure difference that causes the rotor to spin.

The rotor is connected to a shaft, which transfers the rotational energy to a generator housed within the nacelle, the turbine's main body. The nacelle sits atop a tall tower, elevating the rotor to capture stronger and more consistent winds found at higher altitudes. Inside the nacelle, the generator converts the mechanical energy from the spinning rotor into electrical energy. This process involves the use of electromagnetic induction, where the rotation of the shaft causes magnets to move past coils of wire, generating an electric current.

One of the key components of a wind turbine is the gearbox, which is responsible for increasing the rotational speed of the shaft before it reaches the generator. Wind turbines operate at relatively low rotational speeds, but generators require higher speeds to produce electricity efficiently. The gearbox bridges this gap by using a series of gears to amplify the rotor's speed, ensuring optimal performance of the generator. However, some modern turbines use direct-drive technology, eliminating the need for a gearbox and reducing mechanical complexity.

The tower of a wind turbine is another critical element of its design. Towers are typically constructed from steel or concrete and can reach heights of over 100 meters. The height of the tower is a crucial factor in a turbine's ability to capture wind energy, as wind speeds generally increase with altitude. Taller towers allow turbines to access more powerful winds, enhancing their energy output. The tower also provides structural support, anchoring the turbine securely to the ground and withstanding the forces exerted by the wind.

Wind turbines are equipped with a yaw system, which allows them to rotate horizontally to face the wind. This system ensures that the rotor is always optimally aligned with the wind direction, maximizing energy capture. Sensors on the turbine continuously monitor wind direction and speed, sending signals to the yaw motor to adjust the turbine's orientation as needed. This dynamic adjustment is essential for maintaining efficiency and preventing damage from turbulent winds.

The control system of a wind turbine plays a vital role in its functionality, overseeing the operation and safety of the entire structure. This system monitors various parameters, such as wind speed, rotor speed, and power output, to ensure the turbine operates within safe limits. In high wind conditions, the control system can activate the braking mechanism to slow down or stop the rotor, preventing damage from excessive speeds. Additionally, the control system can adjust the pitch of the blades, changing their angle relative to the wind to optimize performance and protect the turbine.

Wind turbines are designed to operate in a wide range of environmental conditions, from gentle breezes to powerful gusts. However, they have a cut-in wind speed, the minimum speed at which they begin generating electricity, and a cut-out wind speed, the maximum speed at which they can safely operate. Beyond the cut-out speed, turbines are shut down to prevent damage. This operational range ensures that turbines can harness wind energy efficiently while maintaining safety and reliability.

The placement of wind turbines is a critical consideration in their design and functionality. Wind farms, which consist of multiple turbines, are strategically located in areas with consistent and strong winds, such as coastal regions, open plains, and hilltops. The spacing between turbines is also important, as it prevents turbulence from affecting the performance of neighboring turbines. Proper siting and spacing maximize the energy output of a wind farm and minimize environmental impacts.

Wind turbine technology continues to evolve, with ongoing research and development focused on improving efficiency, reducing costs, and minimizing environmental impacts. Innovations such as larger rotor diameters, taller towers, and advanced materials are enhancing the performance of wind turbines, making them more competitive with traditional energy sources. Additionally, offshore wind farms are emerging as a promising frontier, taking advantage of the stronger and more consistent winds found over open water.

In summary, wind turbines are complex machines that embody the principles of aerodynamics, mechanics, and electrical engineering. Their design and functionality are the result of meticulous planning and innovation, enabling them to harness the power of the wind and contribute to a sustainable energy future. As technology advances and the demand for clean energy grows, wind turbines will continue to play a pivotal role in the transition to renewable energy sources.

Onshore and Offshore Wind Projects

Wind energy has emerged as a cornerstone of the renewable energy sector, with onshore and offshore wind projects playing pivotal roles in harnessing the power of the wind. Each type of project offers unique advantages and challenges, shaped by geographical, technical, and economic factors. Understanding these distinctions is crucial for anyone interested in the development and implementation of wind energy solutions.

Onshore wind projects are the most common form of wind energy development, characterized by wind turbines installed on land. These projects are typically located in areas with consistent wind patterns, such as open plains, hilltops, and coastal regions. The accessibility of onshore sites makes them relatively straightforward to develop, as they can be connected to existing infrastructure and transportation networks. This ease of access also facilitates the construction and maintenance of wind turbines, reducing overall project costs.

One of the primary advantages of onshore wind projects is their cost-effectiveness. The technology and processes involved in onshore wind development are well-established, leading to lower capital and operational expenses compared to other renewable energy sources. Onshore wind farms can be deployed quickly and efficiently, providing a reliable source of clean energy at a competitive price. Additionally, onshore projects can contribute to local economies by creating jobs and generating tax revenue.

However, onshore wind projects are not without challenges. One of the most significant obstacles is securing suitable land for development. Land use conflicts can arise, particularly in densely populated areas or regions with competing interests, such as agriculture or conservation. Careful site selection and stakeholder engagement are essential to address these concerns and ensure the successful implementation of onshore wind projects. Additionally, the visual and noise impacts of wind turbines can lead to opposition from local communities, necessitating effective communication and mitigation strategies.

Offshore wind projects, on the other hand, involve the installation of wind turbines in bodies of water, typically on the continental shelf. These projects take advantage of the stronger and more consistent winds found over open water, resulting in higher energy yields compared to onshore installations. Offshore wind farms can be located further from populated areas, reducing the visual and noise impacts associated with onshore projects.

The potential for large-scale energy production is a significant advantage of offshore wind projects. The vast expanses of open water provide ample space for the deployment of numerous turbines, allowing for the generation of substantial amounts of electricity. Offshore wind farms can also be situated closer to major coastal cities, reducing transmission losses and providing a reliable energy source for urban centers.

Despite these benefits, offshore wind projects face unique challenges. The harsh marine environment presents technical and logistical difficulties, requiring specialized equipment and expertise for installation and maintenance. The construction of offshore wind farms involves complex engineering and significant capital investment, leading to higher costs compared to onshore projects. Additionally, the integration of offshore wind energy into existing grids can be challenging, necessitating the development of new transmission infrastructure.

Environmental considerations are also a critical factor in the development of offshore wind projects. The potential impact on marine ecosystems and wildlife must be carefully assessed and mitigated to ensure the sustainability of these projects. Environmental impact assessments and stakeholder consultations are essential components of the planning process, helping to identify and address potential issues before construction begins.

The choice between onshore and offshore wind projects depends on a variety of factors, including geographical location, wind resources, and economic considerations. In some cases, a combination of both onshore and offshore projects may be the most effective strategy for maximizing wind energy potential.

Hybrid approaches can leverage the strengths of each type of project, providing a balanced and resilient energy supply.

Technological advancements are continually shaping the landscape of wind energy development. Innovations in turbine design, materials, and installation techniques are enhancing the efficiency and cost-effectiveness of both onshore and offshore projects. Floating wind turbines, for example, are an emerging technology that allows for the deployment of wind farms in deeper waters, expanding the potential for offshore development.

The integration of wind energy with other renewable sources, such as solar and energy storage, is another area of growth. Hybrid renewable energy systems can provide a more stable and reliable energy supply, balancing the variability of wind and solar resources. These integrated solutions can enhance grid stability and resilience, contributing to a more sustainable and secure energy future.

In conclusion, onshore and offshore wind projects each offer distinct advantages and challenges, shaped by their unique characteristics and contexts. By understanding these differences and leveraging the strengths of each type of project, stakeholders can develop effective strategies for harnessing wind energy. As technology continues to advance and the demand for clean energy grows, wind projects will play an increasingly important role in the transition to a sustainable energy future.

Technological Innovations in Wind Energy

Wind energy has long been a cornerstone of the renewable energy sector, and technological innovations continue to propel it forward, enhancing efficiency, reducing costs, and expanding its potential. These advancements are transforming the way wind energy is harnessed, making it a more viable and attractive option for meeting global energy needs. From turbine design to energy storage solutions, the landscape of wind energy is evolving rapidly, driven by cutting-edge technology and innovative thinking.

One of the most significant areas of innovation in wind energy is turbine design. Modern wind turbines are marvels of engineering, incorporating advanced materials and aerodynamic principles to maximize energy capture. The development of larger and more efficient rotor blades has been a key focus, as longer blades can sweep a larger area and capture more wind energy. These blades are often constructed from lightweight composite materials, such as carbon fiber, which provide the necessary strength and durability while minimizing weight.

The introduction of variable-speed turbines has also been a game-changer for wind energy. Unlike traditional fixed-speed turbines, variable speed models can adjust their rotational speed to match wind conditions, optimizing energy capture and reducing mechanical stress. This flexibility enhances the efficiency and lifespan of the turbines, making them more cost-effective over time. Additionally, advanced control systems have been developed to monitor and adjust turbine performance in real-time, ensuring optimal operation and minimizing downtime.

Offshore wind energy has seen remarkable technological advancements, particularly with the development of floating wind turbines. Traditional offshore turbines are anchored to the seabed, limiting their deployment to shallow waters. Floating turbines, however, can be installed in deeper waters, where wind speeds are typically higher and more consistent. This opens up vast new areas for wind energy development, particularly in regions with deep coastal waters. Floating turbines are anchored to the seabed using mooring lines, allowing them to remain stable while harnessing the powerful offshore winds.

Energy storage solutions are another critical area of innovation in wind energy. The intermittent nature of wind power presents challenges for grid stability, as energy production can fluctuate with changing wind conditions. To address this, advanced energy storage systems are being integrated with wind farms, allowing excess energy to be stored and used when needed. Battery storage technology has made significant strides, with lithium-ion batteries leading the way due to their high energy density and declining costs. These batteries can store large amounts of energy in a compact form, providing a reliable backup for wind power.

Hydrogen storage is an emerging solution that holds great promise for wind energy. By using wind power to electrolyze water, hydrogen gas can be produced and stored for later use. This hydrogen can then be converted back into electricity through fuel cells or used directly as a clean fuel source. Hydrogen storage offers the advantage of long-term energy storage, as hydrogen can be stored indefinitely without loss of energy. This makes it an attractive option for balancing seasonal variations in wind energy production.

The integration of wind energy with smart grid technology is another area of innovation that is enhancing the efficiency and reliability of wind power. Smart grids use digital communication and automation to optimize the distribution and consumption of electricity. By integrating wind farms with smart grids, energy can be managed more efficiently, ensuring that wind power is used when it is most needed. This not only enhances the reliability of wind systems but also contributes to grid stability and reduces the need for fossil fuel-based backup power.

Predictive maintenance is a technological advancement that is revolutionizing the operation and maintenance of wind turbines. By using sensors and data analytics, operators can monitor the condition of turbines in real-time and predict when maintenance is needed. This proactive approach reduces downtime and maintenance costs, as issues can be addressed before they lead to significant failures. Predictive maintenance also extends the lifespan of turbines, maximizing their energy output and return on investment.

The use of artificial intelligence and machine learning is also making its mark on the wind energy sector. These technologies are being used to optimize turbine performance, predict wind patterns, and improve energy forecasting. By analyzing vast amounts of data, AI algorithms can identify patterns and trends that would be difficult for humans to discern, leading to more accurate predictions and better decision-making. This enhances the efficiency and reliability of wind energy systems, making them more competitive with traditional energy sources.

As wind energy technology continues to evolve, it is poised to play an increasingly important role in the transition to renewable energy. The ability to harness wind power more efficiently and reliably is crucial for overcoming the limitations of traditional energy sources and ensuring a sustainable energy future. By embracing these technological innovations, we can unlock the full potential of wind energy, paving the way for a cleaner and more sustainable world.

Environmental and Economic Considerations

The transition to renewable energy sources is driven by a dual imperative: the need to mitigate environmental impacts and the desire to harness economic opportunities. Wind energy, as a leading renewable source, offers a compelling case for both environmental and economic benefits. However, understanding the nuances of these considerations is crucial for making informed decisions about wind energy projects.

From an environmental perspective, wind energy is a clean and sustainable resource that produces no greenhouse gas emissions during operation. This makes it a powerful tool in the fight against climate change, as it can significantly reduce the carbon footprint associated with electricity generation. By replacing fossil fuel-based power plants with wind farms, we can decrease air pollution and contribute to cleaner air and healthier ecosystems. The environmental benefits of wind energy extend beyond carbon reduction, as it also helps conserve water resources. Unlike conventional power plants, which require large amounts of water for cooling, wind turbines use no water during operation, reducing the strain on water supplies, particularly in arid regions.

Despite these advantages, wind energy projects must be carefully planned and managed to minimize potential environmental impacts. One of the primary concerns is the effect of wind turbines on wildlife, particularly birds and bats. Collisions with turbine blades can pose a threat to these animals, leading to calls for mitigation measures. Strategies such as siting turbines away from migratory paths, using radar technology to detect approaching birds, and implementing curtailment strategies during peak migration periods can help reduce the risk to wildlife. Additionally, ongoing research and monitoring are essential to understanding and addressing the ecological impacts of wind energy.

The visual and noise impacts of wind turbines are other environmental considerations that can influence public perception and acceptance. The presence of large turbines on the landscape can alter the visual character of an area, leading to opposition from local communities. Noise generated by the rotation of turbine blades can also be a concern, particularly in quiet rural areas. To address these issues, developers can engage with communities early in the planning process, incorporating feedback and exploring design options that minimize visual and noise impacts. This collaborative approach can foster greater acceptance and support for wind energy projects.

On the economic front, wind energy presents a range of opportunities and challenges. One of the most significant economic benefits is the potential for job creation. The wind energy sector has become a major employer, providing jobs in manufacturing, installation, maintenance, and research. As the industry continues to grow, so does the demand for skilled workers, contributing to economic development and job creation in both urban and rural areas. Wind energy projects can also generate tax revenue for local governments, supporting public services and infrastructure development.

The cost of wind energy has decreased significantly over the past decade, making it one of the most competitive sources of electricity. Technological advancements, economies of scale, and increased competition have driven down costs, allowing wind energy to compete with traditional fossil fuels. This cost-effectiveness is a key driver of wind energy adoption, as it provides a reliable and affordable source of electricity. Additionally, wind energy can offer price stability, as it is not subject to the volatility of fossil fuel markets. This can provide economic security for consumers and businesses, protecting them from fluctuating energy prices.

However, the economic viability of wind energy projects depends on several factors, including location, wind resources, and access to transmission infrastructure. The siting of wind farms is a critical consideration, as areas with strong and consistent winds are more likely to yield higher energy outputs and better returns on investment. Proximity to transmission lines is also important, as it affects the cost and feasibility of connecting wind farms to the grid. In some cases, the development of new transmission infrastructure may be necessary to accommodate wind energy, adding to project costs.

Government policies and incentives play a crucial role in shaping the economic landscape of wind energy. Supportive policies, such as tax credits, feed-in tariffs, and renewable energy mandates, can enhance the financial attractiveness of wind projects and encourage investment. Conversely, regulatory barriers and policy uncertainty can hinder development and deter investors. Policymakers have a vital role in creating a stable and supportive environment for wind energy, ensuring that it can thrive and contribute to a sustainable energy future.

The integration of wind energy into existing energy systems presents both opportunities and challenges. Wind power can enhance grid stability and resilience by diversifying the energy mix and reducing reliance on fossil fuels. However, the intermittent nature of wind energy requires careful management to ensure a reliable energy supply. Energy storage solutions, such as batteries and pumped hydro storage, can help balance supply and demand, storing excess energy for use during periods of low wind. Additionally, the development of smart grid technology can optimize the distribution and consumption of wind energy, enhancing efficiency and reliability.

In conclusion, wind energy offers significant environmental and economic benefits, making it a key component of the transition to renewable energy. By understanding and addressing the environmental and economic considerations associated with wind energy projects, stakeholders can make informed decisions that maximize benefits and minimize impacts. As technology continues to advance and the demand for clean energy grows, wind energy will play an increasingly important role in shaping a sustainable and prosperous future.

Policy and Regulatory Support

Navigating the landscape of policy and regulatory support is crucial for the successful deployment and expansion of wind energy projects. Governments and regulatory bodies play a pivotal role in shaping the renewable energy sector, providing the framework within which wind energy can thrive. Understanding the intricacies of policy and regulation is essential for stakeholders seeking to capitalize on the opportunities presented by wind energy.

At the heart of policy support for wind energy are renewable energy targets and mandates. Many countries have set ambitious goals for increasing the share of renewable energy in their overall energy mix, driven by the need to reduce greenhouse gas emissions and combat climate change. These targets provide a clear signal to investors and developers, encouraging the growth of wind energy projects. By establishing long-term commitments to renewable energy, governments can create a stable and predictable environment that fosters investment and innovation.

Financial incentives are another critical component of policy support for wind energy. These incentives can take various forms, including tax credits, grants, and subsidies, all designed to reduce the financial burden of developing and operating wind projects. Production tax credits (PTCs) and investment tax credits (ITCs) are common mechanisms used to incentivize wind energy development. PTCs provide a per-kilowatt-hour tax credit for electricity generated by wind turbines, while ITCs offer a tax credit based on the capital investment in wind projects. These incentives can significantly enhance the economic viability of wind energy, attracting investment and accelerating deployment.

Feed-in tariffs (FITs) are another policy tool used to support wind energy. FITs guarantee a fixed price for electricity generated from renewable sources, providing a stable revenue stream for wind energy producers. This price certainty reduces financial risk and encourages investment in wind projects. By offering long-term contracts at attractive rates, FITs can drive the expansion of wind energy and facilitate the transition to a low-carbon energy system.

Net metering policies also play a role in promoting wind energy, particularly for small-scale and distributed wind projects. Net metering allows consumers who generate their own electricity from wind turbines to feed excess power back into the grid, receiving credit for the energy they contribute. This arrangement can reduce electricity bills and provide an incentive for individuals and businesses to invest in wind energy systems. By enabling consumers to become active participants in the energy market, net metering policies can support the growth of decentralized wind energy.

Regulatory frameworks are essential for ensuring the safe and efficient operation of wind energy projects. These frameworks encompass a range of issues, from permitting and land use to grid integration and environmental protection. Streamlining the permitting process is a key regulatory challenge, as lengthy and complex procedures can delay project development and increase costs. By simplifying and standardizing permitting requirements, governments can facilitate the timely deployment of wind energy projects.

Land use regulations are another important consideration, as they determine where wind projects can be sited. Zoning laws and land use planning can impact the availability of suitable sites for wind farms, particularly in densely populated or environmentally sensitive areas. Engaging with local communities and stakeholders is crucial for addressing land use concerns and securing support for wind projects. By balancing the needs of different stakeholders, regulatory frameworks can ensure that wind energy development is both sustainable and socially acceptable.

Grid integration is a critical regulatory issue, as the intermittent nature of wind energy requires careful management to maintain grid stability. Regulatory bodies must establish rules and standards for connecting wind farms to the grid, ensuring that they can operate efficiently and reliably. This may involve upgrading grid infrastructure, implementing smart grid technology, and developing energy storage solutions to balance supply and demand. By addressing these challenges, regulators can facilitate the integration of wind energy into the broader energy system.

Environmental regulations are also a key consideration for wind energy projects, as they aim to minimize the impact on ecosystems and wildlife. Environmental impact assessments (EIAs) are typically required for wind projects, evaluating potential effects on habitats, species, and landscapes. By identifying and mitigating environmental risks, regulatory frameworks can ensure that wind energy development is compatible with conservation goals. This may involve implementing measures to protect birds and bats, such as adjusting turbine placement or using technology to detect and deter wildlife.

International cooperation and harmonization of policies can further support the growth of wind energy. By aligning regulatory frameworks and sharing best practices, countries can create a more cohesive and efficient global market for wind energy. This cooperation can facilitate cross-border projects, enhance supply chain efficiency, and drive technological innovation. International organizations and agreements, such as the International Renewable Energy Agency (IRENA) and the Paris Agreement, play a vital role in fostering collaboration and advancing the global transition to renewable energy.

In summary, policy and regulatory support are fundamental to the success of wind energy projects. By providing financial incentives, streamlining regulatory processes, and addressing environmental and grid integration challenges, governments and regulatory bodies can create a conducive environment for wind energy development. As the demand for clean energy continues to grow, effective policy and regulation will be essential for unlocking the full potential of wind energy and achieving a sustainable energy future.

Types of Hydropower Systems

Hydropower has long been a cornerstone of renewable energy, harnessing the power of moving water to generate electricity. Its versatility and reliability make it a vital component of the global energy mix. Understanding the different types of hydropower systems is essential for appreciating their unique characteristics and applications. Each system is designed to take advantage of specific water resources and geographical features, offering a range of solutions for energy generation.

The most common type of hydropower system is the conventional dam-based or impoundment system. These systems involve the construction of a dam across a river, creating a reservoir of stored water. The potential energy of the stored water is converted into kinetic energy as it flows through turbines, generating electricity. Dam-based systems are highly efficient and can produce large amounts of electricity, making them suitable for meeting the energy demands of urban and industrial areas. They also offer the advantage of water storage, providing a reliable supply of electricity even during periods of low rainfall.

However, the construction of large dams can have significant environmental and social impacts. The creation of reservoirs can lead to the flooding of vast areas, displacing communities and disrupting ecosystems. Careful planning and mitigation measures are essential to minimize these impacts and ensure the sustainability of dam-based hydropower projects. Additionally, the regulation of water flow can affect downstream ecosystems, requiring careful management to balance energy generation with environmental conservation.

Run-of-river hydropower systems offer an alternative to dam-based systems, with a lower environmental footprint. These systems generate electricity by diverting a portion of a river's flow through a channel or penstock to drive turbines. Unlike dam-based systems, run-of-river projects do not require large reservoirs, reducing the impact on land and ecosystems. They are typically smaller in scale and can be deployed in a variety of settings, from remote rural areas to urban environments.

The intermittent nature of run-of-river systems can be a challenge, as electricity generation is directly tied to river flow. Seasonal variations in water flow can lead to fluctuations in energy output, requiring careful management and integration with other energy sources. Despite these challenges, run-of-river systems offer a flexible and sustainable solution for harnessing hydropower, particularly in regions with abundant river resources.

Pumped storage hydropower is another type of system that plays a crucial role in balancing energy supply and demand. These systems consist of two reservoirs at different elevations, with water being pumped from the lower reservoir to the upper reservoir during periods of low electricity demand. During peak demand, the stored water is released back to the lower reservoir, generating electricity as it flows through turbines. Pumped storage systems act as large-scale batteries, providing a reliable source of energy storage and grid stability.

The ability to store and release energy on demand makes pumped storage systems an invaluable asset for integrating renewable energy sources, such as wind and solar, into the grid.

By smoothing out fluctuations in energy supply, these systems enhance grid resilience and reliability. However, the construction of pumped storage facilities can be capital-intensive and requires suitable geographical conditions, such as natural elevation differences or existing reservoirs.

Micro-hydropower systems represent a smaller-scale approach to hydropower generation, suitable for individual homes, farms, or small communities. These systems typically generate less than 100 kilowatts of electricity and can be installed in small streams or rivers. Micro-hydropower offers a decentralized energy solution, providing reliable and sustainable electricity to remote or off-grid locations. The simplicity and low cost of micro-hydropower systems make them an attractive option for rural electrification and community-based energy projects.

The environmental impact of micro-hydropower is generally minimal, as these systems do not require large infrastructure or significant alterations to watercourses. However, careful site selection and design are essential to ensure that micro-hydropower projects do not disrupt local ecosystems or water availability. By engaging with local communities and stakeholders, developers can ensure that micro-hydropower projects are both sustainable and socially acceptable.

Hybrid hydropower systems combine different types of hydropower technologies or integrate hydropower with other renewable energy sources. These systems offer the potential to optimize energy generation and enhance grid stability by leveraging the strengths of multiple technologies. For example, a hybrid system might combine run-of-river hydropower with solar panels, providing a more consistent and reliable energy supply. By diversifying energy sources, hybrid systems can reduce reliance on fossil fuels and contribute to a more sustainable energy future.

The integration of hydropower with other renewable energy sources is an area of growing interest and innovation. By combining the strengths of different technologies, hybrid systems can provide a more resilient and adaptable energy solution. This approach can enhance the efficiency and reliability of renewable energy systems, making them more competitive with traditional energy sources.

In conclusion, the diversity of hydropower systems offers a range of solutions for harnessing the power of water. From large-scale dam-based systems to small-scale micro-hydropower, each type of system has its unique advantages and challenges. By understanding these differences and selecting the appropriate technology for specific contexts, stakeholders can maximize the benefits of hydropower and contribute to a sustainable energy future. As technology continues to advance and the demand for clean energy grows, hydropower will remain a vital component of the global energy landscape.

Large-Scale vs. Small-Scale Projects

The world of renewable energy is vast and varied, with projects ranging from sprawling installations that power entire cities to small-scale systems designed for individual homes or communities. Understanding the differences between large-scale and small-scale projects is crucial for anyone interested in the renewable energy sector, as each type offers unique benefits and challenges. By examining these distinctions, we can better appreciate the diverse approaches to harnessing clean energy and the roles they play in shaping a sustainable future.

Large-scale renewable energy projects, often referred to as utility-scale projects, are designed to generate significant amounts of electricity, typically feeding directly into the grid. These projects include massive wind farms, solar parks, and hydropower dams, each capable of producing hundreds or even thousands of megawatts of power. The primary advantage of large-scale projects is their ability to meet the energy demands of urban and industrial areas, providing a reliable and consistent supply of electricity. By generating power on a grand scale, these projects can achieve economies of scale, reducing the cost per unit of electricity and making renewable energy more competitive with traditional fossil fuels.

The development of large-scale projects requires substantial investment and planning, often involving complex engineering and construction processes. Securing suitable land is a critical consideration, as these projects require vast areas to accommodate the necessary infrastructure. For example, a large wind farm may span several square miles, while a solar park might cover hundreds of acres. The siting of these projects must take into account factors such as proximity to transmission lines, environmental impact, and community acceptance. Engaging with local stakeholders and conducting thorough environmental assessments are essential steps in the planning process, ensuring that projects are both sustainable and socially responsible.

One of the challenges associated with large-scale projects is their potential impact on the environment and local communities. The construction of massive infrastructure can disrupt ecosystems, alter landscapes, and displace wildlife. To mitigate these impacts, developers must implement strategies such as habitat restoration, wildlife corridors, and careful site selection. Additionally, large-scale projects can face opposition from local communities concerned about visual and noise impacts, land use conflicts, and changes to the local economy. Effective communication and collaboration with stakeholders can help address these concerns and build support for renewable energy initiatives.

In contrast, small-scale renewable energy projects are designed to serve individual homes, businesses, or communities, often operating independently of the grid. These projects include rooftop solar panels, small wind turbines, and micro-hydropower systems, each capable of generating a modest amount of electricity. The primary advantage of small-scale projects is their ability to provide decentralized energy solutions, empowering individuals and communities to generate their own clean power. By reducing reliance on centralized energy systems, small-scale projects can enhance energy security, resilience, and independence.

The development of small-scale projects is typically less capital-intensive and complex than large-scale projects, making them accessible to a wider range of stakeholders. Homeowners, businesses, and community groups can invest in renewable energy systems tailored to their specific needs and resources. The installation of small-scale systems often requires less land and infrastructure, minimizing environmental impact and simplifying the permitting process. Additionally, small-scale projects can be deployed more quickly and flexibly, adapting to changing energy needs and technological advancements.

One of the challenges associated with small-scale projects is their limited capacity to generate electricity, which may not be sufficient to meet all energy demands. To address this, small-scale systems can be integrated with energy storage solutions, such as batteries, to store excess power for use during periods of low generation. Additionally, small-scale projects can be combined with other renewable energy sources, creating hybrid systems that provide a more consistent and reliable energy supply. By leveraging the strengths of multiple technologies, small-scale projects can enhance their efficiency and effectiveness.

The economic viability of small-scale projects depends on factors such as location, available resources, and government incentives. In regions with abundant sunlight or wind, small-scale systems can generate significant amounts of electricity, providing a quick return on investment. Government policies, such as tax credits, rebates, and net metering, can further enhance the financial attractiveness of small-scale projects, encouraging adoption and investment. By creating a supportive policy environment, governments can empower individuals and communities to embrace renewable energy and contribute to a sustainable energy future.

The integration of large-scale and small-scale projects is an emerging trend in the renewable energy sector, offering the potential to optimize energy generation and distribution. By combining centralized and decentralized systems, stakeholders can create a more resilient and adaptable energy network, capable of meeting diverse energy needs. This integrated approach can enhance grid stability, reduce transmission losses, and provide a more balanced energy supply. By leveraging the strengths of both large-scale and small-scale projects, the renewable energy sector can maximize its potential and drive the transition to a low-carbon economy.

In conclusion, large-scale and small-scale renewable energy projects each offer unique advantages and challenges, shaped by their distinct characteristics and contexts. By understanding these differences and selecting the appropriate approach for specific situations, stakeholders can harness the full potential of renewable energy and contribute to a sustainable energy future. As technology continues to advance and the demand for clean energy grows, both large-scale and small-scale projects will play vital roles in shaping the global energy landscape.

Environmental Impact and Mitigation Strategies

Harnessing renewable energy sources is a critical step toward a sustainable future, yet it is essential to recognize and address the environmental impacts associated with these projects. While renewable energy systems, such as wind, solar, and hydropower, offer significant environmental benefits over fossil fuels, they are not without their own ecological footprints. Understanding these impacts and implementing effective mitigation strategies is crucial for ensuring that renewable energy development is both sustainable and environmentally responsible.

Wind energy, for instance, is celebrated for its low carbon emissions and minimal water usage. However, the installation and operation of wind turbines can have notable effects on local ecosystems and wildlife. One of the primary concerns is the risk of bird and bat collisions with turbine blades. These collisions can result in significant mortality rates for certain species, particularly those that migrate or hunt at night. To mitigate this impact, developers can employ strategies such as careful site selection, avoiding areas with high bird and bat activity, and using radar technology to detect and deter wildlife. Additionally, turbine designs that minimize blade movement during low wind conditions can reduce collision risks.

The visual and noise impacts of wind farms are other environmental considerations that can influence public perception and acceptance. The presence of large turbines can alter the visual landscape, leading to concerns about aesthetic impacts. Noise generated by the rotation of turbine blades can also be a concern, particularly in quiet rural areas. To address these issues, developers can engage with communities early in the planning process, incorporating feedback and exploring design options that minimize visual and noise impacts. This collaborative approach can foster greater acceptance and support for wind energy projects.

Solar energy, with its ability to generate electricity without emissions, is another cornerstone of renewable energy. However, large-scale solar installations require significant land use, which can lead to habitat loss and fragmentation. The construction of solar farms can disrupt local ecosystems, affecting plant and animal species. To mitigate these impacts, developers can prioritize the use of previously disturbed or degraded lands, such as brownfields or abandoned agricultural sites, for solar installations. Additionally, integrating solar panels with agricultural activities, a practice known as agrivoltaics, can allow for dual land use, preserving habitat while generating clean energy.

The production and disposal of solar panels also present environmental challenges. The manufacturing process involves the use of hazardous materials, and the disposal of panels at the end of their life cycle can contribute to electronic waste. To address these issues, the solar industry is increasingly focusing on recycling and sustainable manufacturing practices. By developing efficient recycling processes and using eco-friendly materials, the industry can reduce its environmental footprint and promote a circular economy.

Hydropower, as one of the oldest and most established forms of renewable energy, offers reliable and consistent electricity generation. However, the construction of large dams and reservoirs can have profound environmental and social impacts. The creation of reservoirs can lead to the flooding of vast areas, displacing communities and disrupting ecosystems. The regulation of water flow can also affect downstream ecosystems, altering habitats and impacting fish populations. To mitigate these impacts, developers can implement measures such as fish ladders and bypass systems to facilitate fish migration, as well as habitat restoration projects to support affected ecosystems.

Run-of-river hydropower systems offer a lower-impact alternative to traditional dam-based systems, as they do not require large reservoirs. These systems generate electricity by diverting a portion of a river's flow through turbines, minimizing land use and habitat disruption. However, they can still affect river ecosystems, particularly if water diversion reduces flow levels. To mitigate these impacts, developers can implement flow management strategies that maintain ecological flow requirements, ensuring that river ecosystems remain healthy and resilient.

The integration of renewable energy systems with energy storage solutions is another strategy for minimizing environmental impacts. Energy storage technologies, such as batteries and pumped storage, can help balance supply and demand, reducing the need for additional infrastructure and minimizing land use. By storing excess energy for use during periods of low generation, these systems enhance the efficiency and reliability of renewable energy, reducing the overall environmental footprint.

Community engagement and stakeholder involvement are critical components of effective environmental impact mitigation. By involving local communities in the planning and decision-making process, developers can identify and address concerns, build trust, and foster support for renewable energy projects. Transparent communication and collaboration with stakeholders can lead to more informed and sustainable project outcomes, ensuring that environmental and social considerations are integrated into every stage of development.

Policy and regulatory frameworks also play a vital role in guiding environmental impact mitigation efforts. Governments and regulatory bodies can establish standards and guidelines for environmental assessments, ensuring that renewable energy projects are developed in a responsible and sustainable manner. By setting clear expectations and requirements for environmental protection, policymakers can drive innovation and best practices in the renewable energy sector.

In conclusion, while renewable energy systems offer significant environmental benefits, it is essential to recognize and address their potential impacts. By implementing effective mitigation strategies, engaging with communities, and adhering to robust policy frameworks, stakeholders can ensure that renewable energy development is both sustainable and environmentally responsible. As the demand for clean energy continues to grow, these efforts will be crucial for balancing the need for renewable energy with the imperative to protect and preserve our natural world.

Integration into National Grids

The integration of renewable energy sources into national grids is a complex yet essential process for transitioning to a sustainable energy future. As countries strive to reduce their carbon footprints and increase the share of renewables in their energy mix, the challenge of effectively incorporating these variable energy sources into existing grid infrastructure becomes increasingly important. Understanding the intricacies of grid integration is crucial for ensuring a reliable and resilient energy supply.

One of the primary challenges of integrating renewable energy into national grids is the intermittent nature of sources like wind and solar power. Unlike traditional fossil fuel-based power plants, which can provide a consistent and controllable output, renewable energy generation is subject to fluctuations in weather and daylight. This variability can lead to imbalances between supply and demand, posing a challenge for grid operators tasked with maintaining stability and reliability.

To address this challenge, grid operators employ a range of strategies and technologies designed to balance supply and demand in real-time. One such strategy is the use of advanced forecasting techniques to predict renewable energy generation based on weather patterns and other factors. By accurately forecasting the output of wind and solar farms, grid operators can better plan for fluctuations and adjust the dispatch of other power sources accordingly. This proactive approach helps to minimize the impact of variability on grid stability.

Energy storage solutions are another critical component of grid integration, providing a means to store excess renewable energy for use during periods of low generation. Technologies such as batteries, pumped hydro storage, and flywheels offer the ability to smooth out fluctuations in energy supply, enhancing grid resilience and reliability. By storing surplus energy when generation exceeds demand and releasing it when needed, these systems help to balance the grid and reduce reliance on fossil fuel-based backup power.

Demand response programs also play a vital role in integrating renewables into national grids. These programs incentivize consumers to adjust their energy usage in response to grid conditions, helping to balance supply and demand. By shifting energy consumption to periods of high renewable generation, demand response programs can reduce the need for additional infrastructure and enhance the efficiency of the grid. This flexible approach to energy management empowers consumers to play an active role in supporting the transition to renewable energy.

The development of smart grid technology is another key factor in facilitating the integration of renewables. Smart grids use advanced communication and control systems to optimize the distribution and consumption of electricity, enabling more efficient and reliable grid operation. By providing real-time data on energy usage and generation, smart grids allow for more precise management of energy flows, reducing losses and enhancing grid stability. The deployment of smart meters, sensors, and automated controls is transforming the way grids operate, paving the way for a more sustainable and resilient energy system.

Grid infrastructure upgrades are often necessary to accommodate the increased penetration of renewable energy. This may involve reinforcing transmission lines, expanding grid capacity, and developing new interconnections between regions. By enhancing the physical infrastructure of the grid, countries can improve the flow of electricity and reduce bottlenecks, facilitating the integration of renewables. Additionally, the development of regional and international grid interconnections can enable the sharing of renewable resources across borders, enhancing energy security and resilience.

Policy and regulatory frameworks play a crucial role in supporting the integration of renewables into national grids. Governments and regulatory bodies can establish standards and incentives that encourage the development of grid-friendly renewable energy projects. By setting clear guidelines for grid connection and operation, policymakers can drive innovation and best practices in the renewable energy sector. Supportive policies, such as feed-in tariffs and renewable portfolio standards, can also enhance the financial attractiveness of renewable energy projects, encouraging investment and deployment.

The integration of renewables into national grids also requires collaboration and coordination among a wide range of stakeholders, including grid operators, energy producers, policymakers, and consumers. By working together, these stakeholders can develop and implement strategies that address the technical, economic, and regulatory challenges of grid integration. This collaborative approach is essential for ensuring that renewable energy can be effectively and efficiently incorporated into the energy system.

Public engagement and education are also important components of successful grid integration. By raising awareness of the benefits and challenges of renewable energy, stakeholders can build public support for the transition to a sustainable energy future. Engaging with communities and consumers can help to address concerns and misconceptions, fostering a more informed and supportive environment for renewable energy development.

In conclusion, the integration of renewable energy into national grids is a multifaceted challenge that requires a combination of technological, regulatory, and collaborative solutions. By leveraging advanced forecasting, energy storage, demand response, and smart grid technologies, stakeholders can enhance the resilience and reliability of the grid. Policy and regulatory support, along with public engagement and collaboration, are essential for driving the transition to a sustainable energy future. As the demand for clean energy continues to grow, the successful integration of renewables into national grids will be a critical factor in achieving a low-carbon economy.

Future Prospects and Technological Advances

The renewable energy landscape is evolving at an unprecedented pace, driven by technological advances and a growing commitment to sustainability. As the world seeks to transition away from fossil fuels, the future prospects for renewable energy are both promising and transformative. Understanding the potential of emerging technologies and innovations is crucial for anyone interested in the future of energy.

One of the most exciting areas of development is in solar energy, where technological advances are making solar power more efficient and accessible than ever before. Innovations in photovoltaic (PV) technology, such as the development of perovskite solar cells, are pushing the boundaries of efficiency and cost-effectiveness. Perovskite cells offer the potential for higher efficiency rates and lower production costs compared to traditional silicon-based cells. Their flexibility and lightweight nature also open up new possibilities for integration into a variety of surfaces, from building facades to wearable devices.

In addition to advancements in PV technology, solar thermal energy is gaining traction as a viable option for large-scale power generation. Concentrated solar power (CSP) systems use mirrors or lenses to focus sunlight onto a small area, generating heat that can be used to produce electricity. Recent innovations in thermal storage materials and system design are enhancing the efficiency and reliability of CSP systems, making them a competitive option for baseload power generation.

Wind energy is another area where technological advances are driving growth and innovation. The development of larger and more efficient wind turbines is enabling the capture of more energy from the wind, reducing the cost of wind power and expanding its potential applications. Offshore wind farms, in particular, are benefiting from these advances, with floating turbine technology allowing for deployment in deeper waters where wind resources are abundant. This opens up vast new areas for wind energy development, particularly in regions with limited land availability.

Energy storage is a critical component of the renewable energy future, providing the means to balance supply and demand and enhance grid stability. Advances in battery technology, such as the development of solid-state batteries and flow batteries, are increasing the capacity and efficiency of energy storage systems. These innovations are reducing costs and improving the performance of batteries, making them more viable for large-scale applications. Additionally, the integration of energy storage with renewable energy systems is enabling more flexible and reliable energy solutions, supporting the transition to a low-carbon energy system.

Hydrogen is emerging as a key player in the future energy landscape, offering a versatile and clean energy carrier that can complement renewable energy sources. Advances in electrolysis technology are making it more efficient and cost-effective to produce hydrogen from renewable electricity. This green hydrogen can be used for a variety of applications, from powering fuel cells in vehicles to providing industrial heat and energy storage. The development of hydrogen infrastructure, including pipelines and refueling stations, is critical for realizing the full potential of hydrogen as a clean energy solution.

The integration of digital technologies is transforming the way renewable energy systems are managed and operated. The rise of the Internet of Things (IoT), artificial intelligence, and blockchain technology is enabling more efficient and transparent energy management. Smart grids, powered by these digital technologies, allow for real-time monitoring and control of energy flows, optimizing the distribution and consumption of electricity. This digital transformation is enhancing the efficiency and reliability of renewable energy systems, paving the way for a more sustainable and resilient energy future.

The role of policy and regulation in shaping the future of renewable energy cannot be overstated. Governments and regulatory bodies play a crucial role in creating a supportive environment for innovation and investment in renewable energy technologies. By setting ambitious renewable energy targets, providing financial incentives, and establishing clear regulatory frameworks, policymakers can drive the development and deployment of new technologies. International cooperation and collaboration are also essential for advancing the global transition to renewable energy, as countries share best practices and work together to overcome common challenges.

Public engagement and education are vital for fostering support for renewable energy and encouraging the adoption of new technologies. By raising awareness of the benefits and potential of renewable energy, stakeholders can build public support for the transition to a sustainable energy future. Engaging with communities and consumers can help to address concerns and misconceptions, fostering a more informed and supportive environment for renewable energy development.

In conclusion, the future prospects for renewable energy are bright, driven by technological advances and a growing commitment to sustainability. From solar and wind energy to hydrogen and digital technologies, the innovations shaping the renewable energy landscape are transforming the way we generate and consume energy. By embracing these advances and fostering a supportive policy and regulatory environment, stakeholders can unlock the full potential of renewable energy and drive the transition to a low-carbon energy system. As the demand for clean energy continues to grow, the future of renewable energy holds the promise of a more sustainable and resilient world.

Chapter 4: Biomass and Bioenergy: Nature's Renewable Resource

Understanding Biomass Conversion Processes

Biomass conversion processes are at the heart of transforming organic materials into usable energy, offering a renewable alternative to fossil fuels. Understanding these processes is essential for anyone interested in the potential of biomass as a sustainable energy source. Biomass, which includes plant materials, agricultural residues, and organic waste, can be converted into energy through various methods, each with its unique advantages and applications.

One of the most common biomass conversion processes is combustion, where organic material is burned to produce heat and power. This process is similar to traditional fossil fuel combustion but uses renewable biomass as the fuel source. Combustion is a straightforward and well-established technology, making it a popular choice for generating electricity and heat. Biomass combustion systems can range from small-scale stoves and boilers for residential heating to large power plants that supply electricity to the grid. The key to efficient biomass combustion lies in optimizing the combustion conditions, such as temperature and air supply, to maximize energy output and minimize emissions.

Gasification is another biomass conversion process that involves heating organic material in a low-oxygen environment to produce a combustible gas known as syngas. Syngas, composed mainly of carbon monoxide and hydrogen, can be used to generate electricity, produce heat, or serve as a feedstock for chemical synthesis. Gasification offers several advantages over direct combustion, including higher efficiency and the ability to produce a cleaner-burning fuel. The process can handle a wide range of biomass feedstocks, from wood chips to agricultural residues, making it a versatile option for biomass conversion.

Pyrolysis is a thermal decomposition process that occurs in the absence of oxygen, breaking down biomass into a mixture of solid, liquid, and gaseous products. The primary products of pyrolysis are biochar, bio-oil, and syngas. Biochar, a carbon-rich solid, can be used as a soil amendment to improve soil fertility and sequester carbon. Bio-oil, a liquid product, can be upgraded to produce transportation fuels or used as a feedstock for chemical production. Pyrolysis offers the potential for producing a range of valuable products from biomass, making it an attractive option for integrated biorefineries.

Anaerobic digestion is a biological process that converts organic material into biogas through the action of microorganisms in the absence of oxygen. Biogas, composed mainly of methane and carbon dioxide, can be used to generate electricity, produce heat, or serve as a vehicle fuel. Anaerobic digestion is particularly well-suited for wet biomass feedstocks, such as animal manure, food waste, and sewage sludge. The process not only produces renewable energy but also reduces waste volumes and produces nutrient-rich digestate, which can be used as a fertilizer. Anaerobic digestion systems can be implemented at various scales, from small on-farm digesters to large centralized facilities.

Fermentation is another biological process used to convert biomass into biofuels, such as ethanol and butanol. The process involves the breakdown of sugars by microorganisms, typically yeast or bacteria, to produce alcohols. Ethanol fermentation is a well-established technology, with corn and sugarcane being the primary feedstocks. However, advances in biotechnology are enabling the use of lignocellulosic biomass, such as agricultural residues and woody materials, for ethanol production. This second-generation biofuel technology offers the potential to expand the range of feedstocks and reduce competition with food crops.

The choice of biomass conversion process depends on several factors, including the type of biomass feedstock, the desired end products, and the scale of the operation. Each process has its unique advantages and challenges, and selecting the appropriate technology requires careful consideration of these factors. For example, combustion and gasification are well-suited for dry biomass feedstocks, while anaerobic digestion is ideal for wet materials. Pyrolysis and fermentation offer the potential for producing a range of valuable products, but may require more complex processing and upgrading steps.

The integration of biomass conversion processes into existing energy systems presents both opportunities and challenges. Biomass can be co-fired with coal in existing power plants, reducing greenhouse gas emissions and extending the life of these facilities. However, the variability in biomass feedstock quality and supply can pose challenges for consistent operation. Developing robust supply chains and feedstock management strategies is essential for ensuring a reliable and sustainable biomass energy system.

Environmental considerations are also important when evaluating biomass conversion processes. While biomass is a renewable resource, its production and conversion can have environmental impacts, such as land use changes, water consumption, and emissions. Implementing sustainable biomass sourcing practices, such as using agricultural residues and waste materials, can help mitigate these impacts. Additionally, optimizing conversion processes to maximize efficiency and minimize emissions is crucial for reducing the environmental footprint of biomass energy.

Policy and regulatory frameworks play a critical role in supporting the development and deployment of biomass conversion technologies. Governments can provide incentives for biomass energy projects, establish standards for sustainable biomass sourcing, and support research and development efforts. By creating a supportive policy environment, policymakers can drive innovation and investment in biomass energy, contributing to a more sustainable energy future.

Public engagement and education are also important for fostering support for biomass energy and encouraging the adoption of new technologies. By raising awareness of the benefits and potential of biomass, stakeholders can build public support for the transition to a sustainable energy future. Engaging with communities and consumers can help address concerns and misconceptions, fostering a more informed and supportive environment for biomass energy development.

In conclusion, understanding biomass conversion processes is essential for harnessing the potential of biomass as a renewable energy source. From combustion and gasification to anaerobic digestion and fermentation, each process offers unique advantages and applications. By selecting the appropriate technology and implementing sustainable practices, stakeholders can unlock the full potential of biomass energy and contribute to a low-carbon energy system. As the demand for clean energy continues to grow, biomass conversion processes will play a vital role in shaping the future of renewable energy.

Biofuels: Production and Applications

Biofuels have emerged as a promising alternative to traditional fossil fuels, offering a renewable and potentially more sustainable source of energy. Derived from organic materials, biofuels can be produced from a variety of feedstocks, including crops, agricultural residues, and waste materials. Understanding the production processes and applications of biofuels is essential for anyone interested in the future of renewable energy.

The production of biofuels begins with the selection of suitable feedstocks, which can vary widely depending on regional availability and economic considerations. First-generation biofuels are typically derived from food crops such as corn, sugarcane, and soybeans. These crops are processed to produce ethanol and biodiesel, the two most common types of biofuels. Ethanol is produced through the fermentation of sugars, while biodiesel is produced through the transesterification of oils and fats. While first-generation biofuels have been successful in reducing greenhouse gas emissions compared to fossil fuels, they have faced criticism for competing with food production and contributing to land use changes.

To address these concerns, second-generation biofuels have been developed, utilizing non-food feedstocks such as agricultural residues, woody biomass, and dedicated energy crops like switchgrass and miscanthus. These feedstocks are often more abundant and less expensive than food crops, and their use can help mitigate the food-versus-fuel debate. The production of second-generation biofuels involves more complex processes, such as the breakdown of lignocellulosic biomass into fermentable sugars through pretreatment and enzymatic hydrolysis. These sugars are then fermented to produce ethanol or other biofuels. Advances in biotechnology and process engineering are continually improving the efficiency and cost-effectiveness of second-generation biofuel production.

Third-generation biofuels, derived from algae and other microorganisms, represent the cutting edge of biofuel technology. Algae have the potential to produce high yields of oil and biomass, making them an attractive feedstock for biodiesel and other biofuels. Algae can be cultivated in a variety of environments, including wastewater and non-arable land, reducing competition with food production. The production of algal biofuels involves the cultivation of algae in photobioreactors or open ponds, followed by the extraction and conversion of oils into biodiesel or other fuels. While still in the early stages of commercialization, third-generation biofuels offer the potential for sustainable and scalable biofuel production.

The applications of biofuels are diverse, spanning transportation, power generation, and industrial processes. In the transportation sector, biofuels are used as a substitute for gasoline and diesel, reducing greenhouse gas emissions and dependence on fossil fuels. Ethanol is commonly blended with gasoline to produce E10 or E85, while biodiesel can be used in its pure form or blended with petroleum diesel. The use of biofuels in transportation is supported by government mandates and incentives, which have driven the growth of biofuel markets worldwide.

In addition to transportation, biofuels can be used for power generation, providing a renewable alternative to coal and natural gas. Biomass power plants can co-fire biofuels with traditional fuels, reducing emissions and extending the life of existing infrastructure. Biofuels can also be used in combined heat and power (CHP) systems, which generate electricity and heat simultaneously, improving overall energy efficiency. The use of biofuels in power generation is particularly attractive in regions with abundant biomass resources and supportive policy frameworks.

Industrial applications of biofuels include the production of chemicals, plastics, and other materials. Bio-based chemicals and materials offer a renewable alternative to petroleum-based products, reducing the environmental impact of industrial processes. The development of biorefineries, which integrate the production of biofuels and bio-based products, is driving innovation and value creation in the bioeconomy. By utilizing all components of biomass feedstocks, biorefineries can maximize resource efficiency and economic viability.

The environmental benefits of biofuels are significant, but their production and use also present challenges. The cultivation of biofuel feedstocks can have environmental impacts, such as land use changes, water consumption, and biodiversity loss. Implementing sustainable agricultural practices, such as crop rotation and conservation tillage, can help mitigate these impacts. Additionally, optimizing biofuel production processes to maximize efficiency and minimize emissions is crucial for reducing the environmental footprint of biofuels.

Policy and regulatory frameworks play a critical role in supporting the development and deployment of biofuels. Governments can provide incentives for biofuel production and use, establish standards for sustainable feedstock sourcing, and support research and development efforts. By creating a supportive policy environment, policymakers can drive innovation and investment in biofuels, contributing to a more sustainable energy future.

Public engagement and education are also important for fostering support for biofuels and encouraging the adoption of new technologies. By raising awareness of the benefits and potential of biofuels, stakeholders can build public support for the transition to a sustainable energy future. Engaging with communities and consumers can help address concerns and misconceptions, fostering a more informed and supportive environment for biofuel development.

In conclusion, biofuels offer a renewable and potentially more sustainable alternative to traditional fossil fuels, with diverse applications in transportation, power generation, and industry. From first-generation biofuels derived from food crops to advanced third-generation biofuels from algae, the production and use of biofuels are continually evolving. By selecting suitable feedstocks, optimizing production processes, and implementing sustainable practices, stakeholders can unlock the full potential of biofuels and contribute to a low-carbon energy system. As the demand for clean energy continues to grow, biofuels will play a vital role in shaping the future of renewable energy.

Environmental and Economic Impacts

The transition to renewable energy sources is not only a technological shift but also a profound transformation with significant environmental and economic implications. As societies grapple with the challenges of climate change and resource depletion, understanding the impacts of renewable energy adoption becomes crucial for informed decision-making and sustainable development.

From an environmental perspective, renewable energy sources offer a cleaner and more sustainable alternative to fossil fuels. The most immediate benefit is the reduction in greenhouse gas emissions, which are the primary drivers of climate change. By replacing coal, oil, and natural gas with wind, solar, hydro, and other renewables, countries can significantly decrease their carbon footprints. This shift is essential for meeting international climate targets and mitigating the adverse effects of global warming, such as rising sea levels, extreme weather events, and biodiversity loss.

Beyond carbon emissions, renewable energy technologies also contribute to improved air quality. Traditional fossil fuel combustion releases pollutants like sulfur dioxide, nitrogen oxides, and particulate matter, which can lead to respiratory illnesses and other health problems. By reducing reliance on these fuels, renewable energy can help decrease air pollution and its associated health costs. This improvement in air quality is particularly beneficial in urban areas, where pollution levels are often highest and populations are most vulnerable.

The environmental benefits of renewable energy extend to water resources as well. Conventional power plants, especially those using coal and nuclear energy, require significant amounts of water for cooling and other processes. This demand can strain local water supplies, particularly in arid regions or during droughts. In contrast, most renewable energy technologies, such as wind and solar, have minimal water requirements, reducing the pressure on water resources and contributing to more sustainable water management.

However, the deployment of renewable energy technologies is not without its environmental challenges. The production and installation of solar panels, wind turbines, and other infrastructure can have ecological impacts, such as habitat disruption and resource extraction. For instance, the mining of rare earth elements used in some renewable technologies can lead to environmental degradation if not managed responsibly. Additionally, large-scale renewable projects, like hydroelectric dams, can alter ecosystems and displace communities. Addressing these challenges requires careful planning, environmental assessments, and the implementation of best practices to minimize negative impacts.

On the economic front, the transition to renewable energy presents both opportunities and challenges. One of the most significant economic benefits is the potential for job creation. The renewable energy sector is labor-intensive, requiring a diverse workforce for manufacturing, installation, maintenance, and research. As the industry grows, it can create millions of jobs worldwide, contributing to economic development and reducing unemployment. Moreover, many of these jobs are in local communities, supporting regional economies and fostering economic resilience.

Renewable energy also offers the potential for energy independence and security. By harnessing domestic energy resources, countries can reduce their reliance on imported fossil fuels, which are subject to price volatility and geopolitical tensions. This shift can enhance national security and stabilize energy prices, benefiting consumers and businesses alike. Furthermore, the decentralized nature of many renewable energy systems, such as rooftop solar panels, can increase the resilience of energy infrastructure, reducing vulnerability to disruptions and natural disasters.

The economic viability of renewable energy has improved significantly in recent years, driven by technological advancements and economies of scale. The cost of solar and wind power has plummeted, making them competitive with, or even cheaper than, fossil fuels in many regions. This cost reduction has spurred investment and accelerated the deployment of renewable energy projects. However, the transition also requires substantial upfront investment in infrastructure, research, and development. Governments and private investors must work together to mobilize the necessary capital and create a supportive policy environment that encourages innovation and deployment.

Despite the economic benefits, the transition to renewable energy can pose challenges for certain industries and communities. The decline of fossil fuel industries, such as coal mining and oil extraction, can lead to job losses and economic disruption in regions dependent on these sectors. Managing this transition requires proactive policies and support for affected workers and communities, such as retraining programs, economic diversification initiatives, and social safety nets. By addressing these challenges, policymakers can ensure a just and equitable transition that leaves no one behind.

The integration of renewable energy into existing energy systems also presents economic challenges. The variability of some renewable sources, like wind and solar, requires investments in grid infrastructure, energy storage, and demand management to ensure a reliable energy supply. These investments can be costly, but they are essential for maximizing the benefits of renewable energy and maintaining grid stability. Innovative financing mechanisms, such as green bonds and public-private partnerships, can help mobilize the necessary resources and spread the costs over time.

Public engagement and education are critical for fostering support for renewable energy and addressing economic and environmental concerns. By raising awareness of the benefits and challenges of renewable energy, stakeholders can build public support for the transition and encourage the adoption of new technologies. Engaging with communities and consumers can help address concerns and misconceptions, fostering a more informed and supportive environment for renewable energy development.

In conclusion, the environmental and economic impacts of renewable energy are profound and multifaceted. By reducing greenhouse gas emissions, improving air and water quality, and creating jobs, renewable energy offers a pathway to a more sustainable and resilient future. However, the transition also presents challenges that require careful planning, investment, and collaboration among stakeholders. By addressing these challenges and leveraging the opportunities, societies can unlock the full potential of renewable energy and contribute to a low-carbon, sustainable energy system. As the demand for clean energy continues to grow, understanding the environmental and economic impacts of renewable energy will be essential for shaping the future of energy.

Challenges in Scaling Biomass Solutions

Scaling biomass solutions presents a complex array of challenges that must be navigated to fully harness the potential of this renewable energy source. As the world seeks sustainable alternatives to fossil fuels, biomass offers a promising avenue, yet its expansion is fraught with technical, economic, and environmental hurdles. Understanding these challenges is crucial for developing effective strategies to overcome them and ensure the successful integration of biomass into the global energy mix.

One of the primary challenges in scaling biomass solutions is the availability and sustainability of feedstocks. Biomass feedstocks, which include agricultural residues, forestry by-products, and dedicated energy crops, must be sourced in a manner that does not compete with food production or lead to deforestation and habitat loss. The variability in feedstock supply, influenced by seasonal changes and geographic distribution, can complicate logistics and supply chain management. Developing robust supply chains that ensure a consistent and sustainable feedstock supply is essential for scaling biomass solutions. This requires collaboration among farmers, foresters, and energy producers, as well as the implementation of sustainable agricultural and forestry practices.

The conversion of biomass into energy or biofuels involves complex processes that can be costly and technologically challenging. Technologies such as combustion, gasification, and anaerobic digestion each have their own set of technical requirements and limitations. For instance, the efficiency of biomass combustion systems can be affected by the moisture content and composition of the feedstock, while gasification processes require precise control of temperature and pressure conditions. Scaling these technologies requires ongoing research and development to improve efficiency, reduce costs, and enhance the reliability of biomass conversion systems. Investment in pilot projects and demonstration plants can help bridge the gap between laboratory research and commercial-scale deployment.

Economic viability is another significant challenge in scaling biomass solutions. While the cost of biomass feedstocks can be competitive with fossil fuels, the capital investment required for biomass conversion facilities can be substantial. Additionally, the economic feasibility of biomass projects is often influenced by market conditions, such as the price of competing energy sources and government incentives. Policymakers play a crucial role in creating a supportive economic environment for biomass by providing subsidies, tax incentives, and other financial mechanisms that encourage investment in biomass technologies. Developing innovative financing models, such as public-private partnerships and green bonds, can also help mobilize the necessary capital for scaling biomass solutions.

The integration of biomass into existing energy systems presents technical and regulatory challenges. Biomass power plants must be compatible with existing grid infrastructure, and the intermittent nature of some biomass feedstocks can pose challenges for grid stability. Developing advanced grid management systems and energy storage solutions is essential for ensuring a reliable energy supply. Regulatory frameworks must also be adapted to accommodate the unique characteristics of biomass energy, including standards for emissions, feedstock sustainability, and land use. Policymakers must work closely with industry stakeholders to develop regulations that support the growth of biomass while protecting environmental and social interests.

Environmental considerations are paramount when scaling biomass solutions. While biomass is a renewable resource, its production and use can have environmental impacts, such as land use changes, water consumption, and emissions. Implementing sustainable sourcing practices, such as using agricultural residues and waste materials, can help mitigate these impacts. Additionally, optimizing conversion processes to maximize efficiency and minimize emissions is crucial for reducing the environmental footprint of biomass energy. Life cycle assessments can provide valuable insights into the environmental impacts of biomass projects, guiding decision-making and promoting best practices.

Public perception and acceptance of biomass solutions can also influence their scalability. Misconceptions about the environmental impacts and sustainability of biomass can hinder public support and investment. Engaging with communities and stakeholders through education and outreach initiatives is essential for building trust and fostering acceptance of biomass technologies. Transparent communication about the benefits and challenges of biomass, as well as the measures being taken to address environmental and social concerns, can help build public confidence and support for biomass projects.

International cooperation and collaboration are vital for overcoming the challenges of scaling biomass solutions. Sharing best practices, research findings, and technological innovations can accelerate the development and deployment of biomass technologies. Collaborative efforts can also help address global challenges, such as climate change and energy security, by promoting the widespread adoption of sustainable biomass solutions. International organizations, governments, and industry stakeholders must work together to create a global framework that supports the growth of biomass and facilitates the exchange of knowledge and resources.

In conclusion, scaling biomass solutions requires a multifaceted approach that addresses technical, economic, environmental, and social challenges. By developing sustainable feedstock supply chains, advancing conversion technologies, and creating supportive policy and regulatory frameworks, stakeholders can unlock the full potential of biomass as a renewable energy source. Public engagement and international collaboration are also essential for fostering acceptance and accelerating the deployment of biomass solutions. As the demand for clean energy continues to grow, overcoming the challenges of scaling biomass will be crucial for achieving a sustainable and resilient energy future.

Innovations and Future Directions

Innovation is the lifeblood of progress, and in the realm of renewable energy, it is the driving force propelling us toward a sustainable future. As the world grapples with the pressing need to transition away from fossil fuels, the role of innovation in shaping the future of energy cannot be overstated. From cutting-edge technologies to novel approaches in policy and infrastructure, the landscape of renewable energy is evolving at an unprecedented pace, offering exciting possibilities and new directions.

One of the most promising areas of innovation lies in the development of advanced materials and technologies for energy capture and storage. Solar energy, for instance, has seen remarkable advancements with the introduction of perovskite solar cells. These materials offer the potential for higher efficiency and lower production costs compared to traditional silicon-based cells. Researchers are exploring ways to enhance the stability and scalability of perovskite cells, aiming to bring them to market as a viable alternative. Similarly, innovations in wind turbine design, such as bladeless turbines and floating offshore platforms, are expanding the potential for harnessing wind energy in diverse environments.

Energy storage is another critical frontier in the renewable energy sector. The intermittent nature of solar and wind power necessitates efficient storage solutions to ensure a reliable energy supply. Advances in battery technology, particularly in lithium-ion and solid-state batteries, are paving the way for more efficient and longer-lasting storage systems. Researchers are also exploring alternative storage methods, such as flow batteries and hydrogen fuel cells, which offer the potential for large-scale energy storage and grid stability. These innovations are crucial for integrating renewable energy into existing infrastructure and reducing reliance on fossil fuels.

The digital revolution is playing a transformative role in the renewable energy sector, with innovations in data analytics, artificial intelligence, and the Internet of Things (IoT) driving efficiency and optimization. Smart grids, which leverage real-time data and advanced analytics, are enabling more efficient energy distribution and consumption. These grids can dynamically balance supply and demand, integrate distributed energy resources, and enhance grid resilience. IoT devices and sensors are also being deployed to monitor and optimize energy use in homes, businesses, and industrial settings, contributing to greater energy efficiency and cost savings.

Policy and regulatory innovations are equally important in shaping the future of renewable energy. Governments around the world are implementing policies to incentivize the adoption of clean energy technologies, such as feed-in tariffs, tax credits, and renewable portfolio standards. These policies are designed to create a favorable market environment for renewable energy, encouraging investment and driving down costs. Additionally, international cooperation and agreements, such as the Paris Agreement, are fostering global collaboration and commitment to reducing greenhouse gas emissions and transitioning to sustainable energy systems.

The concept of a circular economy is gaining traction as a framework for sustainable development, with implications for the renewable energy sector. In a circular economy, resources are used more efficiently, waste is minimized, and materials are recycled and reused. This approach is being applied to the design and production of renewable energy technologies, such as solar panels and wind turbines, to reduce their environmental impact and extend their lifespan. Innovations in recycling and repurposing materials are helping to close the loop and create more sustainable energy systems.

Community-based renewable energy projects are emerging as a powerful model for local empowerment and sustainable development. These projects, often initiated and managed by local communities, provide an opportunity for individuals to participate in and benefit from the transition to renewable energy. Community solar programs, for example, allow residents to invest in and share the benefits of solar energy, even if they cannot install panels on their own properties. These initiatives not only promote renewable energy adoption but also foster social cohesion and economic resilience.

Looking to the future, the potential for innovation in renewable energy is vast and varied. Emerging technologies, such as artificial photosynthesis and fusion energy, hold the promise of revolutionizing the way we produce and consume energy. Artificial photosynthesis seeks to mimic the natural process of photosynthesis to convert sunlight, water, and carbon dioxide into fuels and chemicals, offering a sustainable and carbon-neutral energy source. Fusion energy, often referred to as the "holy grail" of energy, aims to replicate the processes that power the sun, providing a virtually limitless and clean energy source. While these technologies are still in the experimental stages, they represent exciting possibilities for the future of energy.

The integration of renewable energy into urban environments is another area ripe for innovation. As cities continue to grow and urbanize, the demand for sustainable energy solutions is increasing. Innovations in urban planning and design, such as green roofs, solar facades, and energy-efficient buildings, are helping to create more sustainable and resilient cities. The concept of "smart cities," which leverage technology and data to optimize energy use and reduce emissions, is gaining momentum as a model for sustainable urban development.

Education and workforce development are critical components of the renewable energy transition. As the industry evolves, there is a growing need for skilled workers and professionals who can design, install, and maintain renewable energy systems. Educational institutions and training programs are adapting to meet this demand, offering courses and certifications in renewable energy technologies and sustainable practices. By investing in education and workforce development, we can ensure a pipeline of talent to support the growth and innovation of the renewable energy sector.

In conclusion, the future of renewable energy is bright, driven by a wave of innovation and creativity. From advanced materials and technologies to policy and community initiatives, the renewable energy landscape is evolving rapidly, offering new opportunities and directions. By embracing innovation and collaboration, we can overcome the challenges of the energy transition and build a sustainable and resilient energy future. As we look ahead, the potential for renewable energy to transform our world is limitless, and the journey is just beginning.

Chapter 5: Geothermal Energy: Unlocking Earth's Heat

Geothermal Energy Systems Explained

Geothermal energy, a powerful yet often underappreciated resource, taps into the Earth's internal heat to provide a sustainable and reliable energy source. This form of energy harnesses the thermal energy stored beneath the Earth's crust, offering a clean alternative to fossil fuels. Understanding the intricacies of geothermal energy systems is essential for appreciating their potential and implementing them effectively.

At the heart of geothermal energy is the natural heat generated by the Earth's core, which is primarily the result of radioactive decay and residual heat from the planet's formation. This heat gradually moves towards the surface, creating geothermal reservoirs that can be accessed for energy production. These reservoirs are typically found in regions with high tectonic activity, such as the Pacific Ring of Fire, where the Earth's crust is thinner and more permeable.

Geothermal energy systems can be broadly categorized into three types: dry steam, flash steam, and binary cycle. Each system utilizes different methods to convert geothermal heat into usable energy, depending on the temperature and pressure of the geothermal resource.

Dry steam systems are the simplest and oldest form of geothermal technology. They directly use steam from geothermal reservoirs to drive turbines and generate electricity. This method is highly efficient but requires geothermal fields with naturally occurring steam, which are relatively rare. The Geysers in California, the largest dry steam field in the world, exemplifies this technology's potential, providing a significant portion of the state's renewable energy.

Flash steam systems, the most common type of geothermal power plant, utilize high-pressure hot water from geothermal reservoirs. The hot water is brought to the surface and depressurized, or "flashed," into steam. This steam is then used to drive turbines and generate electricity. Flash steam plants are versatile and can operate with a wide range of geothermal resources, making them suitable for many locations worldwide.

Binary cycle systems represent the latest advancement in geothermal technology, allowing for the exploitation of lower-temperature geothermal resources. In these systems, geothermal fluid is passed through a heat exchanger, where it transfers its heat to a secondary fluid with a lower boiling point. This secondary fluid vaporizes and drives a turbine to generate electricity. Binary cycle plants have minimal emissions and can be used in areas with moderate geothermal resources, expanding the potential for geothermal energy development.

The environmental benefits of geothermal energy are significant. Unlike fossil fuels, geothermal systems produce minimal greenhouse gas emissions, contributing to a reduction in air pollution and climate change mitigation. Additionally, geothermal plants have a small land footprint compared to other renewable energy sources, such as solar and wind farms, making them an attractive option for regions with limited available land.

Despite these advantages, geothermal energy systems face several challenges that must be addressed to maximize their potential. One of the primary challenges is the high upfront cost associated with exploration and drilling. Identifying viable geothermal reservoirs requires extensive geological surveys and test drilling, which can be expensive and time-consuming. However, advancements in exploration technology, such as remote sensing and geophysical imaging, are helping to reduce these costs and improve the success rate of geothermal projects.

Another challenge is the potential for induced seismicity, or human-caused earthquakes, associated with geothermal energy extraction. The injection and withdrawal of fluids in geothermal reservoirs can alter subsurface pressures and trigger seismic events. While most induced seismicity is minor and poses little risk to human safety, it is essential to monitor and manage these activities to minimize potential impacts. Implementing best practices and regulatory frameworks can help mitigate the risks associated with induced seismicity.

Geothermal energy systems also face competition from other renewable energy sources, such as solar and wind, which have seen significant cost reductions and technological advancements in recent years. To remain competitive, the geothermal industry must continue to innovate and improve efficiency. Hybrid systems, which combine geothermal energy with other renewable sources, offer a promising avenue for enhancing performance and reliability. For example, integrating solar thermal energy with geothermal systems can increase overall efficiency and provide a more consistent energy supply.

The potential for direct use applications of geothermal energy extends beyond electricity generation. Geothermal heat can be used for various purposes, including district heating, greenhouse agriculture, aquaculture, and industrial processes. These applications can provide significant energy savings and reduce reliance on fossil fuels. For instance, Iceland, a leader in geothermal energy utilization, uses geothermal heat for district heating, providing warmth to nearly 90% of its population.

Geothermal heat pumps, a technology that utilizes the stable temperatures of the Earth's subsurface, offer another avenue for harnessing geothermal energy. These systems can efficiently heat and cool buildings by transferring heat between the ground and the building. Geothermal heat pumps are highly efficient and can significantly reduce energy consumption and greenhouse gas emissions in residential and commercial buildings.

The future of geothermal energy systems is promising, with ongoing research and development efforts aimed at overcoming existing challenges and expanding the potential for geothermal energy utilization. Enhanced geothermal systems (EGS), which involve creating artificial reservoirs by fracturing hot dry rock, offer the potential to access vast untapped geothermal resources. EGS technology is still in the experimental stage, but successful implementation could revolutionize the geothermal industry and significantly increase the availability of geothermal energy.

International collaboration and knowledge sharing are essential for advancing geothermal energy systems. Countries with extensive geothermal experience, such as Iceland, New Zealand, and the United States, can provide valuable insights and expertise to emerging geothermal markets. Collaborative efforts can accelerate the development and deployment of geothermal technologies, contributing to global energy security and sustainability.

In conclusion, geothermal energy systems offer a sustainable and reliable energy source with significant environmental benefits. By understanding the different types of geothermal systems and addressing the challenges they face, we can unlock the full potential of geothermal energy and contribute to a cleaner, more sustainable energy future. As technology continues to advance and new opportunities emerge, geothermal energy will play an increasingly important role in the global transition to renewable energy.

Global Geothermal Potential and Hotspots

Geothermal energy, a formidable force beneath our feet, holds immense potential to transform the global energy landscape. This renewable resource, derived from the Earth's internal heat, offers a sustainable and reliable alternative to fossil fuels. Understanding the global potential and identifying geothermal hotspots is crucial for harnessing this energy source effectively and integrating it into the broader energy mix.

The Earth's geothermal potential is vast, with estimates suggesting that the heat stored beneath the Earth's crust could supply the world's energy needs for millennia. This potential is not evenly distributed, however, as geothermal resources are concentrated in specific regions where geological conditions are favorable. These regions, often referred to as geothermal hotspots, are characterized by high heat flow, tectonic activity, and the presence of geothermal reservoirs.

The Pacific Ring of Fire, a horseshoe-shaped zone encircling the Pacific Ocean, is one of the most prominent geothermal hotspots. This region is home to numerous active volcanoes and tectonic plate boundaries, creating ideal conditions for geothermal energy production. Countries such as Indonesia, the Philippines, and Japan have significant geothermal resources and are actively developing them to meet their energy needs. Indonesia, in particular, has emerged as a global leader in geothermal energy, with ambitious plans to expand its geothermal capacity and reduce its reliance on fossil fuels.

The East African Rift Valley is another notable geothermal hotspot, offering significant potential for energy development. This geological feature, which stretches from the Red Sea to Mozambique, is characterized by high heat flow and volcanic activity. Countries such as Kenya and Ethiopia have made substantial investments in geothermal energy, recognizing its potential to provide a stable and sustainable energy supply. Kenya's Olkaria Geothermal Plant, one of the largest in Africa, exemplifies the successful harnessing of geothermal resources in the region.

The western United States, particularly the states of California, Nevada, and Utah, is rich in geothermal resources. The Geysers in California, the largest geothermal field in the world, has been a pioneer in geothermal energy production for decades. The region's unique geological conditions, including the presence of fault lines and volcanic activity, create an ideal environment for geothermal development. The United States continues to invest in geothermal research and development, exploring new technologies and expanding its geothermal capacity.

Iceland, a small island nation in the North Atlantic, is renowned for its extensive use of geothermal energy. The country's unique geological setting, located on the Mid-Atlantic Ridge, provides abundant geothermal resources. Iceland has leveraged this natural advantage to become a global leader in renewable energy, with geothermal power supplying a significant portion of its electricity and heating needs. The country's success in harnessing geothermal energy serves as a model for other nations seeking to develop their geothermal potential.

New Zealand, situated on the Pacific Ring of Fire, also boasts significant geothermal resources. The country's geothermal power plants contribute a substantial share of its electricity generation, providing a reliable and sustainable energy source. New Zealand's commitment to renewable energy and its expertise in geothermal technology position it as a key player in the global geothermal landscape.

While these regions represent some of the most prominent geothermal hotspots, geothermal potential exists in many other parts of the world. Central and South America, for example, have significant geothermal resources, with countries like Mexico, Costa Rica, and El Salvador actively developing their geothermal capacity. Europe, too, has untapped geothermal potential, with countries such as Italy, Turkey, and Germany exploring opportunities for geothermal energy production.

The development of geothermal energy is not without its challenges. Identifying viable geothermal reservoirs requires extensive geological surveys and exploration, which can be costly and time-consuming. The high upfront costs associated with drilling and infrastructure development can also pose barriers to entry. However, advancements in exploration technology and drilling techniques are helping to reduce these costs and improve the feasibility of geothermal projects.

Environmental considerations are paramount in the development of geothermal energy. While geothermal systems produce minimal greenhouse gas emissions, they can have localized environmental impacts, such as land subsidence and the release of trace gases. Implementing best practices and regulatory frameworks can help mitigate these impacts and ensure the sustainable development of geothermal resources.

The integration of geothermal energy into existing energy systems requires careful planning and coordination. Geothermal power plants must be compatible with grid infrastructure, and the variability of geothermal resources can pose challenges for grid stability. Developing advanced grid management systems and energy storage solutions is essential for maximizing the benefits of geothermal energy and ensuring a reliable energy supply.

International collaboration and knowledge sharing are vital for unlocking the global geothermal potential. Countries with extensive geothermal experience can provide valuable insights and expertise to emerging geothermal markets. Collaborative efforts can accelerate the development and deployment of geothermal technologies, contributing to global energy security and sustainability.

The future of geothermal energy is promising, with ongoing research and development efforts aimed at overcoming existing challenges and expanding the potential for geothermal energy utilization. Enhanced geothermal systems (EGS), which involve creating artificial reservoirs by fracturing hot dry rock, offer the potential to access vast untapped geothermal resources. EGS technology is still in the experimental stage, but successful implementation could revolutionize the geothermal industry and significantly increase the availability of geothermal energy.

In conclusion, the global geothermal potential is vast and varied, offering a sustainable and reliable energy source for countries around the world. By identifying geothermal hotspots and addressing the challenges associated with geothermal development, we can unlock the full potential of this renewable resource and contribute to a cleaner, more sustainable energy future. As technology continues to advance and new opportunities emerge, geothermal energy will play an increasingly important role in the global transition to renewable energy.

Environmental and Economic Benefits

Harnessing renewable energy sources offers a multitude of environmental and economic benefits, making them an attractive alternative to traditional fossil fuels. As the world grapples with the dual challenges of climate change and energy security, the transition to renewable energy presents a viable solution that addresses both concerns. By exploring the environmental and economic advantages of renewable energy, we can better understand its potential to transform our energy systems and create a more sustainable future.

One of the most significant environmental benefits of renewable energy is its potential to reduce greenhouse gas emissions. Unlike fossil fuels, which release carbon dioxide and other harmful pollutants when burned, renewable energy sources such as solar, wind, and geothermal produce little to no emissions during operation. This reduction in emissions is crucial for mitigating the impacts of climate change and improving air quality. By transitioning to renewable energy, we can decrease our reliance on fossil fuels and significantly reduce our carbon footprint, contributing to global efforts to combat climate change.

Renewable energy also offers the advantage of being a sustainable and inexhaustible resource. Unlike finite fossil fuels, which are subject to depletion and price volatility, renewable energy sources are abundant and naturally replenished. The sun, wind, and Earth's internal heat provide a continuous supply of energy that can be harnessed without depleting natural resources. This sustainability ensures a long-term energy supply that can support economic growth and development without compromising the needs of future generations.

The environmental benefits of renewable energy extend beyond emissions reduction and resource sustainability. Renewable energy projects often have a smaller land footprint compared to fossil fuel extraction and power generation. For example, solar panels can be installed on rooftops or integrated into existing infrastructure, minimizing land use and habitat disruption. Wind farms can coexist with agricultural activities, allowing for dual land use and preserving natural landscapes. Additionally, renewable energy systems typically require less water for operation compared to conventional power plants, reducing the strain on water resources and minimizing the risk of water pollution.

The economic benefits of renewable energy are equally compelling. The renewable energy sector has become a significant driver of job creation and economic growth. As the demand for clean energy technologies increases, so does the need for skilled workers in manufacturing, installation, maintenance, and research and development. The renewable energy industry has already created millions of jobs worldwide, and this trend is expected to continue as more countries invest in renewable energy infrastructure. These jobs often provide higher wages and better working conditions compared to traditional energy sectors, contributing to improved livelihoods and economic resilience.

Renewable energy also offers the potential for energy cost savings and price stability. While the initial investment in renewable energy infrastructure can be substantial, the long-term operational costs are typically lower than those of fossil fuel-based systems. Renewable energy sources have no fuel costs, as they rely on natural resources that are freely available. This can lead to significant savings on energy bills for consumers and businesses. Moreover, renewable energy prices are less susceptible to market fluctuations and geopolitical tensions, providing greater price stability and energy security.

The decentralization of energy production is another economic benefit of renewable energy. Distributed energy systems, such as rooftop solar panels and small-scale wind turbines, allow individuals and communities to generate their own electricity, reducing their dependence on centralized power grids. This decentralization can enhance energy resilience, particularly in remote or underserved areas, by providing a reliable and independent energy supply. It also empowers communities to take control of their energy future, fostering local economic development and self-sufficiency.

Renewable energy can also stimulate innovation and technological advancement. The transition to clean energy requires the development of new technologies and solutions, driving research and development efforts across various sectors. This innovation can lead to the creation of new industries and markets, further boosting economic growth and competitiveness. As countries invest in renewable energy research and development, they position themselves as leaders in the global clean energy transition, attracting investment and fostering international collaboration.

The environmental and economic benefits of renewable energy are interconnected and mutually reinforcing. By reducing emissions and promoting sustainability, renewable energy contributes to a healthier environment, which in turn supports economic prosperity. Clean air and water, preserved ecosystems, and a stable climate are essential for human health and well-being, as well as for the productivity of industries such as agriculture, tourism, and fisheries. By investing in renewable energy, we can create a virtuous cycle of environmental protection and economic growth.

Despite these benefits, the transition to renewable energy is not without challenges. The integration of renewable energy into existing energy systems requires careful planning and investment in infrastructure, such as grid upgrades and energy storage solutions. Policymakers play a crucial role in creating a supportive regulatory environment that encourages renewable energy development and addresses potential barriers. Financial incentives, such as tax credits and subsidies, can help offset the initial costs of renewable energy projects and accelerate their deployment.

Public awareness and acceptance are also critical for the successful adoption of renewable energy. Educating communities about the benefits of renewable energy and addressing concerns about potential impacts, such as noise from wind turbines or land use changes, can foster support and engagement. Community involvement in renewable energy projects, such as cooperative ownership models, can enhance local buy-in and ensure that the benefits are shared equitably.

In conclusion, the environmental and economic benefits of renewable energy make it a compelling choice for a sustainable and prosperous future. By reducing emissions, conserving resources, and creating jobs, renewable energy offers a pathway to address the pressing challenges of climate change and energy security. As we continue to invest in and develop renewable energy technologies, we can unlock their full potential and create a cleaner, more resilient world for generations to come. The journey toward a renewable energy future is not only necessary but also filled with opportunities for innovation, growth, and positive change.

Technological Challenges and Solutions

Navigating the landscape of renewable energy involves overcoming a series of technological challenges that stand in the way of widespread adoption and integration. These challenges, while formidable, are not insurmountable. With innovation and strategic planning, solutions can be developed to address these obstacles, paving the way for a cleaner and more sustainable energy future.

One of the primary technological challenges in renewable energy is the issue of intermittency. Solar and wind power, two of the most prevalent renewable sources, are inherently variable. The sun doesn't always shine, and the wind doesn't always blow, leading to fluctuations in energy production. This variability poses a significant challenge for maintaining a stable and reliable energy supply. To address this, advancements in energy storage technology are crucial. Batteries, particularly lithium-ion and emerging solid-state technologies, are at the forefront of this effort. These storage systems can capture excess energy generated during peak production times and release it when demand is high or production is low, effectively smoothing out the supply curve.

Grid integration is another critical challenge. The existing energy infrastructure was primarily designed for centralized power generation from fossil fuels, not for the decentralized and variable nature of renewable energy. Upgrading the grid to accommodate renewable sources involves implementing smart grid technologies that can dynamically manage energy flows, integrate distributed energy resources, and enhance grid resilience. These smart grids rely on real-time data and advanced analytics to optimize energy distribution and ensure stability. Additionally, microgrids, which are localized energy systems that can operate independently or in conjunction with the main grid, offer a solution for integrating renewable energy in remote or underserved areas.

The efficiency of renewable energy technologies is another area where significant improvements are needed. While solar panels and wind turbines have become more efficient over the years, there is still room for enhancement. Research and development efforts are focused on developing new materials and designs that can increase the efficiency of energy capture and conversion. For instance, perovskite solar cells, a promising alternative to traditional silicon-based cells, offer the potential for higher efficiency and lower production costs. Similarly, innovations in wind turbine design, such as vertical-axis turbines and bladeless models, aim to improve performance and expand the range of viable installation sites.

The high upfront costs associated with renewable energy projects can also be a barrier to adoption. While the long-term operational costs of renewable energy are typically lower than those of fossil fuels, the initial investment required for infrastructure development can be substantial. To overcome this challenge, financial mechanisms such as green bonds, power purchase agreements, and government incentives can play a crucial role. These tools can help offset the initial costs and make renewable energy projects more financially viable. Additionally, economies of scale and technological advancements are driving down costs, making renewable energy increasingly competitive with traditional energy sources.

Another challenge lies in the environmental impact of renewable energy technologies. While they offer significant environmental benefits compared to fossil fuels, renewable energy systems can still have localized impacts. For example, the production and disposal of solar panels and wind turbine blades can generate waste and require the use of rare materials. To address these concerns, the concept of a circular economy is being applied to renewable energy technologies. This approach focuses on designing products for longevity, recyclability, and minimal environmental impact. Innovations in recycling and repurposing materials are helping to close the loop and create more sustainable energy systems.

Public perception and acceptance are also critical factors in the successful deployment of renewable energy technologies. Concerns about visual impact, noise, and land use can lead to opposition from local communities. Engaging with communities early in the planning process and addressing their concerns through transparent communication and participatory decision-making can foster support and acceptance. Community ownership models, where local residents have a stake in renewable energy projects, can also enhance buy-in and ensure that the benefits are shared equitably.

The rapid pace of technological advancement presents both challenges and opportunities for the renewable energy sector. Keeping up with the latest developments and integrating new technologies into existing systems requires continuous learning and adaptation. Collaboration between industry, academia, and government is essential for driving innovation and ensuring that new technologies are effectively deployed. International cooperation and knowledge sharing can also accelerate the development and adoption of renewable energy solutions, contributing to global energy security and sustainability.

The role of policy and regulation cannot be overlooked in addressing technological challenges. Policymakers play a crucial role in creating a supportive environment for renewable energy development. This includes setting ambitious targets, implementing supportive policies, and removing barriers to entry. Regulatory frameworks that encourage innovation, protect the environment, and ensure fair competition are essential for fostering a thriving renewable energy sector.

In conclusion, while the technological challenges facing renewable energy are significant, they are not insurmountable.

Through innovation, collaboration, and strategic planning, solutions can be developed to overcome these obstacles and unlock the full potential of renewable energy. By addressing issues such as intermittency, grid integration, efficiency, and cost, we can pave the way for a cleaner, more sustainable energy future. The journey toward renewable energy is filled with opportunities for growth, innovation, and positive change, and by embracing these challenges, we can create a more resilient and sustainable world for generations to come.